Special Operations Executive

Operational Stores Handbook

English langauage version

ISBN-13: 978-1976343421
ISBN-10: 1976343429

(C) CA Brown, 2017

Photographic reproduction of the cardboard covers of the original handbook. Completely unmarked. Image from CA Brown collection

Introductory

The British Special Operations Executive (SOE) trained, advised, in some cases commanded, and equipped resistance groups fighting the Nazi invaders in mainland Europe from Denmark and Norway in the north, to Yugoslavia and Albania in the Balkans.

Operational stores such as weapons, ammunition, radios and demolitions equipment and explosives were dropped in special parachute containers over secluded fields in the dead of night, and the supplies had to be broken down and distributed, or simply cached all before daylight since word often filtered back to the Germans about mysterious aircraft in the night, and the Gestapo always followed up reports, hoping to capture resistance fighters with a load of guerrilla stores fresh from England.

Each weapon and demolitions stores equipment container contained a small handbook. Under its nondescript plain cardboard covers, the handbook contained sabotage, explosives and weapons instructions as well as a vocabulary of standard SOE operational stores, complete with illustrations of the various devices and their use. The handbook was divided into seven different, colour coded sections, each in a different language - English, French, Dutch, Norwegian, Danish, Polish and German. It is the English section and accompanying illustrations reproduced here.

The instructions were simple and to the point and in SOE's style, were suited for use by untrained or barely trained guerrillas, but were also a valuable refresher for more experienced personnel.

The vocabulary of stores found in the "contents" section of the handbook provided standard equipment item numbers and designations, as well as the designation of the standardised stores cell the items would be found in within the main stores containers. This information greatly simplified the process of requesting stores from London to be parachuted into occupied territory.

Very hard to find, this is, as far as is known, the first time this enigmatic little handbook has ever been printed since the end of the Second World War. It stands as stark testimony to a clandestine war, fought in the shadows, which continues to keep some dark secrets to this day.

CA Brown
September, 2017

FOREWORD.

This pamphlet replaces the old one which has been out of date for some time. It is printed in various languages and all the illustrations are at the end.

Each store has a serial number, and when the name first appears it is printed in heavy type. Subsequently it is not. The table of contents gives a complete list of stores and the cells in which they will be found, together with instructions for ordering standard containers.

The pamphlet is not intended to be a complete guide to the use of the stores, and the contents have been made as brief as possible.

A drawing of each store is given to assist in the identification of items.

TABLE OF CONTENTS.

Item No.		Section in Booklet	Cell in which found	Remarks
1.	Plastic H.E.	A	H 1, 2, 3, 4, 18	
2.	Nobel's 808	A	H 1, 2, 3, 4,	Will be sent only if P.E. is in short supply.
3.	C.E. Primer	A	H 3 4	
4.	Detonator	A	H 5	
5.	Bickford Safety Fuse	A	H 5, 10, 11, 12	
6.	Crimping Tool	A	H 5	
7.	Fusee Matches	A	H 5, 10	
8.	Striker Board	A	H 5, 11, 12	
9.	Copper Tube Igniter	A	H 5	
10.	Pull Switch	A	H 5	
11.	Balloon	A	H 5	
12.	Adhesive Tape	A	H 5, 10, 11, 12, 23, 24	
13.	Cordtex	B	H 1, 2, 3	
14.	Copper Sealing Cap	B	H 5	
15.	Clam	C	H 13	
16.	General Purpose Charge	C	—	Supplies will be sent when available in near future.
17.	Rubberised Fabric	C	H 5	
18.	Bostik	C	H 5	
19.	Rubber Solution	C	H 5	
20.	1 1/2 lb. Charge	C	—	Supplies will be sent when available.
21.	Rail Charge	C, D	H 21, 22	
22.	Magnets	C	—	May not be sent if in short supply.
23.	Fog Signal	D	H 5	
24.	Pencil Time Fuse	E	H 5, 11, 12	
25.	Firepot	F	H 11, 12	Some 1 lb. Incendiaries and 2 oz. Paraffin Incendiaries may still be sent, but the new Incendiary Block (No. 27) will eventually replace these.
26.	2 1/2 lb. Thermit	F	H 11, 12	
27.	Incendiary Block	F	—	
28.	Pocket Time Incendiary	F	H 11, 12	
29.	Instantaneous Fuse	F	H 11, 12	
30.	Limpet	G	H 23, 24	
31.	A.C. Delay	G	H 23, 24	
32.	Spanner	G	H 23, 24	
33.	Luting	G	H 23, 24	Included in A.C. Delay Tin.
34.	Placing Rod	G	H 23, 24	

Item No.		Section in Booklet	Cell in which found	Remarks
35.	Magnetic Holdfast	G	H 23, 24	
36.	Trip Wire		H 5	
37.	Pressure Switch	H	—	Omitted at present while modifications are being made.
38.	Release Switch	H	H 5	
39.	No. 82 Grenade	I	H 10	
40.	Primer Grenade 82	I	H 10	
41.	Special Detonator	I	H 10	
42.	Percussion Fuse (Allways) for G.P.	I	—	For use with General Purpose Charge.
43.	Tyre Burster	I		At present in "C" Type container only.
44.	Detonator for Tyre Burster	I	—	
45.	No. 36 Mills Grenade	I	H 8, 9	
46.	Detonator for No. 36 Mills Grenade	I	H 8, 9	
47.	Abrasive Paste	I	H 13	
48.	Pliers	I	H 13	
49.	Vaseline	I	H 5	
50.	Field Dressings	I	H 8, 9	
51.	Sten S.M.G.	J	H 6, 7	
52.	Pistols	K	H 8, 13	These may be either revolvers or automatics. The latter are in very short supply

The items listed above are packed in standard containers as follows:—

Container Type	Buffer Head	Cell Position and No.					Parachute Bucket	Description Contents
		A.	B.	C.	D.	E.		
H. 1		1	2	3	4	5		H.E. & ancillaries
H. 2		6	6	7	7	7		Sten
H. 3		6	9	7	8	10		Weapon
H. 4		11	11	12	12	12		Incendiary
H. 5		1	2	3	13	5		Sabotage

When ordering a consignment of Containers you need only say "Send X H5 and X H3" for example.

A. HOW TO USE EXPLOSIVES.

The easiest way for small parties of offensively minded men to attack the enemy is by the use of explosives. Immense amount of damage can be caused with a small quantity of supplies.

If the instructions given in this Booklet are followed carefully, even a person with no previous knowledge of explosives will be able to hit the enemy where it will do him most damage.

What is an Explosive?

An explosive is a substance which can be made to produce an explosion. There are many kinds, but those delivered with this pamphlet are the most powerful, and are particularly suitable for cutting metal.

In order that these explosives shall be safe to handle in quantity, they are made as insensitive as possible. They should always be handled carefully, but when in good condition they will not explode if dropped, crushed or set on fire.

How to Fire Explosives.

Since the explosive is made insensitive it requires to be given some shock to set it off. This shock, to increase safety, is sub-divided into two stages:—
   (a) The Primer.
   (b) The Detonator.

The primer consists of a small quantity of explosive, more sensitive than the explosive itself, but less sensitive than the detonator which contains very sensitive explosive, suitably protected by a metal case.

The detonator is initiated by a flame, and in order that this flame can be applied safely by the operator, slow burning safety fuse, which can be lit by a match or other form of igniter, is supplied.

To make an explosive explode or detonate, therefore, the following are required:—
   (1) Flame or spark.
   (2) Safety fuse.
   (3) Detonator.
   (4) Primer.
   (5) Explosive.

1.—The Explosive.

There are two explosives which will normally be delivered:—
   (a) Plastic High Explosive.
   (b) Nobel's 808.

Any other types which may be sent in lieu will have special instructions sent with them.

It should be noted that these instructions apply only to Plastic and 808, and not to explosives which may be found locally.

(a) **PLASTIC H.E.** (No. 1.)   (Diagram 1.)

It is yellow in colour and is supplied in cellophane covered cartridges weighing 112 grammes, or in 224 gramme cartridges in waxed paper. It is sometimes supplied in 2.3 kilo. slabs, or ready made up into charges. (See Section C, sub-section 6).

This is a very powerful explosive and very safe to handle, and it can be moulded easily into any shape with no ill effects to the user. It will not explode if hit by a bullet.

At low temperatures it may be difficult to mould, but this can be got over by placing the cartridges in the trouser pocket until the cartridges are soft.

If the explosive is used in petrol or oil it must be "waterproofed" because it is destroyed by these liquids.

(b) **NOBEL'S 808.** (No. 2.) (Diagram 2.)

It is brownish in colour, and of a rubbery nature, and is in sticks weighing approximately 112 gm. It smells strongly of almonds. This is also a very powerful explosive, and is excellent for cutting metal.

It must be in close contact with the object to be cut, and to ensure this, it is possible to cut the explosive so as to obtain a flat surface—the knife must be used in the same manner as cutting a cake, that is to say, there should be no sawing motion.

It is also possible to mould 808 after it has been warmed for about $\frac{1}{4}-\frac{1}{2}$ hour by placing it under the armpit. It should not be placed in front of a fire. It can also be softened by placing the cartridges in a tin surrounded by boiling water, but the water container should not be placed over a flame. When handling 808, do not handle for long with bare hands or you will get a bad headache.

When storing 808, it should be kept as dry as possible and not allowed to get too hot or too cold.

2.—THE PRIMER. (No. 3.) (Diagram 3.)

This is a small quantity of more sensitive explosive made in a shape like a cork. It has a hole running through the centre to take a detonator (see next paragraph).

The Primer must be in close contact with the main explosive. The detonator must be inserted into the primer hole, but must not be pushed right through—it should stop just short of the end of the hole. (Diagram 10). If the detonator fits loosely, it should be packed with a blade of grass or something similar. The Primer should be kept dry—it has a damp-proof coating and this must not be chipped.

There must always be explosive between the Primer and the target.

3.—THE DETONATOR. (No. 4.) (Diagram 4.)

This is a small tube half filled with very sensitive explosive, and open at one end. It is quite safe to handle, providing normal care is taken. It will give a sufficient shock to detonate a primer. The detonator itself can be fired by a spark.

The inside of the detonator must be protected against damp during storage. Sawdust is sometimes used for packing and this must all be carefully tapped out before use.

4.—BICKFORD SAFETY FUSE. (No. 5.) (Diagram 5.)

This is a slow burning fuse which burns at a rate of 1 cm. per second and when the burning reaches the far end, a spark is produced which will fire a detonator.

Safety fuse can easily be identified by the black powder filling.

Cut a suitable length of safety fuse and insert one end into the detonator, pushing it gently home as far as it will go. Then " crimp " or squeeze the open end of the detonator to prevent the safety fuse pulling out. This can be done either with the **CRIMPING TOOL (No. 6)** or with the teeth. (Diagrams 6, 7, 8 and 9.)

The fuse is waterproof and will burn under water, but cut ends are easily affected by damp. The rolls of safety fuse should be stored in a dry place and before use at least 15 cms. should be cut off the end of each roll in case damp has reached the filling. A further length should then be tested for speed of burning.

5.—Ways of Lighting Safety Fuse.

(a) **FUSEE. (No. 7.)**

The Fusee is a non-flaming match which can be used with an ordinary safety matchbox, or with a **STRIKER BOARD (No. 8.)** It burns at the side and not the end.

The safety fuse should be cut at a slant to expose more of the filling. (Diagram 11).

(b) **Ordinary Match.**

The match-head should be held against the filling and the box drawn across the match. (Diagram 12).

(c) **COPPER TUBE IGNITER. (No. 9.)**

This consists of a small copper tube with a match-head at one end. It should be slipped over the end of the safety fuse and crimped into place. The safety fuse should be cut square. (Diagram 13).

A striker board or safety matchbox is required to ignite it. (Diagram 14).

(d) **THE PULL SWITCH. (No. 10.)**

This is a simple mechanism suitable for igniting fuse and also as a booby trap. It has a spring snout (1) at one end to receive the safety fuse, and at the other end a metal loop (2) which can be pulled. A safety pin (3) prevents the striker inside the device from coming down on to the cap in the base, before it should. To operate the igniter, remove the safety pin (3) and pull on the loop (2).

The use of the igniter as a booby trap will be described later in the pamphlet. (See Section H).

6. Preparing a Simple Charge.

A charge consists of two parts:—
 (a) The Explosive Charge.
 (b) The Firing System.

The explosive charge consists of the required amount of explosive with a primer embedded in it.

The firing system normally consists of the detonator, safety fuse and igniter. These two parts should be kept separate until the explosive charge is in place and ready for firing, when the detonator should be inserted and the igniter lit.

7.—Waterproofing the Firing System. (Diagrams 16 and 17).

It is essential that Copper Tube Igniters should be kept dry and this can be done by using a BALLOON. (No. 11.)

Cut 1 cm. off the Striker Board (1) and push it inside the balloon (2).

Cut off the rubber ring (3) on the end of the balloon. Push the copper tube igniter (4) into the balloon and tape the neck of the balloon tightly to the safety fuse (5), using ADHESIVE TAPE. (No. 12) (6).

Then slip the rubber ring over the balloon, constricting it between the copper tube igniter and the striker board, to prevent the two rubbing together.

To operate the firing system, remove the rubber ring and by manipulating the striker board inside the balloon, strike it across the copper tube igniter.

B. FIRING CHARGES SIMULTANEOUSLY.

When it is required to fire two or more charges simultaneously, they must be linked together. **CORDTEX (No. 13)** is provided for this purpose.

This fuse is an explosive and can only be fired by a detonator—NOT by safety fuse or other igniter.

It can be recognised by its white filling. The outer covering is waterproof but exposed ends are ruined by damp and should always be sealed with a **COPPER SEALING CAP (No. 14.)** It can safely be cut with a knife.

1.—Use of Cordtex.
Cordtex has three main uses:—
(a) For linking several charges together.
(b) for making up charges so that they can be fired in inaccessible places.
(c) As a substitute for a primer.

(a) Linking Charges together.
This is explained fully in sub-section 3 of Section C.

(b) Making up Charges with Cordtex.
This is fully explained in Section C. A simple method is illustrated in **Diagram 20**.

(c) As a substitute for a Primer.
If primers are not available, Cordtex can be used as a substitute. A special knot must be tied in the place normally occupied by the primer. The method of doing this is illustrated in **Diagrams 18 and 19**.

2.—Firing Cordtex.
The detonator or detonators should be taped to the Cordtex with the closed end pointing down the Cordtex to the charge. They should always be taped to the Cordtex at least 10 cms. from the free end.

Diagram 20.
(1) Cordtex.   (2) Detonator.
(3) Tape.   (4) Primer.
(5) Charge.   (6) Sealing Cap.

3.—Precautions to Observe when Using Cordtex.
(a) Leads must never cross (**Diagram 21**)—lead C is wrong and will be cut, charge C not firing.
(b) All leads must go in the same direction (**Diagram 22**). Lead C is incorrect and charge C will not fire.
(c) Acute angles must be avoided.
(5) All ends of Cordtex must be waterproofed with the special copper sealing caps provided, which should be slipped over the end and crimped into place.

C. HOW MUCH EXPLOSIVE TO USE AND HOW TO PREPARE CHARGES.

Experience has shewn that it is essential to concentrate on a small number of charges and to learn what they will do.

The following charges are suitable for a wide range of targets:—
(1) The ½ lb. Clam.
(2) The 1 lb. General Purpose Charge.
(3) The 1½ lb. Charge.
(4) The 3 lb. Charge.
(5) The Rail Charge.

The ½ lb. Clam and the 1 lb. General Purpose Charge are ready made charges; the remainder can be made up from cartridge explosive. The 1½ lb. charge and the Rail Charge are in some cases also ready made, but differ from the others in being flexible and not enclosed in a container.

The ready made 1½ lb. charge and Rail Charge are described at the end of this Section.

1.—THE 1/2 lb. CLAM (224 gm.) (No. 15.) (Diagram 23.)

This consists of a bakelite container (1) filled with explosive; magnets (2) are fitted on the underside. A detonator can be inserted through the slot (3) in the top of the container into the primer embedded in the explosive filling.

When placing the charge, care should be taken to ensure that the explosive portion of the Clam is in contact with the part of the target to be cut.

2.—The 1 lb. GENERAL PURPOSE CHARGE (448 gm.) (No. 16.)
(Diagram 24.)

This charge is enclosed in a metal case (1), down the centre of which runs a hollow metal sleeve sealed at one end with a wooden plug (2). To the plug is attached a spring (3), and when a detonator is inserted into the open end (4) of the metal sleeve this spring prevents the detonator from falling out and maintains it in the correct position in relation to the primers inside the charge. (Diagram 25).

When linking General Purpose Charges with Cordtex, the plug is withdrawn and the detonating fuse is then passed straight through the sleeve. (Diagram 26.)

One end of the charge is fitted with a threaded socket to take a percussion fuse, enabling the charge to be used, in emergency, as a grenade. (See Section H, sub-section 2).

3.—The 1 1/2 lb. Charge.

This charge can be made up either of P.E. or 808. The materials required are:—
(a) Explosive P.E. or 808. 672 grammes. (1).
(b) Two primers. (2).
(c) Cordtex. (3).
(d) RUBBERISED FABRIC or other material. (No. 17.) (4.)
(e) BOSTIK—special sticking solution in tubes. (No. 18.) (5).
RUBBER SOLUTION. (No. 19.)

The charge is made up as follows:—

(i) Mould six cartridges of explosive into a block approximately 5 cms. x 5 cms. x 15 cms. (Diagram 27).

(ii) Cut the block in half down its length. (Diagram 28).

(iii) Cut two lengths of detonating fuse (1), each 1 m. 35 cms. long.

(iv) Thread a primer (2) on each length of fuse, one from the right hand and one from the left hand end with the small ends pointing towards the centre. Secure the primers 2 ft. from their respective ends by means of tape or by making a knot (3) in the detonating fuse so that they will not slide along the fuse. (Diagram 29).

(v) Place the detonating fuse and primers between the two halves of the explosive block so that one primer is at each end of the block with the narrow end pointing inwards. Press the two halves together.

(vi) Re-shape the block of explosive, which now has two primers and double tails at each end, into a block 5.6 cms. x 4.7 cms. x 16.25 cms.

(vii) Wrap the block of explosive in waterproof fabric, cloth or strong paper. Fold the ends over neatly, first cutting a slot or hole in the fabric or paper through which the Cordtex tails can be threaded. Secure the wrapping with Bostik or tape. (Diagram 30).

Where the explosive is likely to be subjected to a high temperature a stiffening of thin cardboard can, with advantage, be placed underneath the wrapping.

(viii) Seal the ends of the Cordtex with Bostik or by wrapping a small piece of tape round each one, or use a Copper Sealing Cap. (4).

4.—The 3 lb. Charge.

The method of preparing this charge is exactly the same as that described for the 1¼ lb. charge, but more explosive is needed. The dimensions should be:—

```
        Length  10¼"  (26.25 cms).
        Height   2¼  ( 5.625 cms.)
        Width    2¼" ( 5.625 cms.)
```

5.—The Rail Charge.

This charge is a double charge linked by Cordtex, designed specifically for use in the derailment of trains. Its application is described under targets.

To make up this charge:—

(i) Cut a length of Cordtex (1) 4½ metres long.

(ii) Take two primers (2), double the Cordex and thread the primers on to one length.

(iii) Tape them into position with a gap of just over 1 metre between the two primers. (Diagram 31).

(iv) Round each primer mould ¾ lb. of Plastic (336 gms.) or 808, into a charge suitable to fit tightly between the top and bottom flanges of a rail. (Diagram 32).

There must be plenty of explosive between the primer and the rail.

(v) Cut a small piece of wood (3) and place on the outside of the charge. (Diagram 33).

(vi) Cover the whole charge with rubberised fabric.

6.—THE 1 1/2 lb. MADE UP CHARGE. (No. 20.)   (Diagram 34.)
This is a charge of P.E. made up in a rubberised fabric cover with the primers arranged as in the General Purpose Charge. It can be used in exactly the same way as the General Purpose Charge and can be cut in half making two ¾ lb. charges, suitable for use as rail charges as well as for other purposes.

7.—THE MADE UP RAIL CHARGE. (No. 21.)
This charge consists of two ¾ lb. charges of explosive in rubber bags, the cartridges being loose inside the wrapping. Special attachments are provided to make fixing easy. These are more fully described in Section D, sub-section 3 (f).

8.—Dividing Charges.
The 1½ lb. Charge described in Section 3 can easily be cut in half to form two separate ¾ lb. charges. (Diagram 35).

It can also be split, by cutting the explosive only, forming two charges which can be slid apart along the Cordtex. The tape on the Cordtex at each end of the charge, however, must be removed. (Diagram 36).

9.—Linking Charges.
The charges described above are of two types, those with double tails of Cordtex and those with single tails. The method of linking is, however, essentially the same.
    (i) Junction Box Linking. (Diagram 37).
    (ii) Straight Main Linking. (Diagram 38).
    (iii) Linking in series. (Diagram 39).

10.—How to Fix a Charge to a Target.
The charge, to do the maximum amount of damage must be in close contact with the target, over the whole area of the charge. If a space cannot be avoided, fill it in with mud or similar material.

Fix the charge securely in place, using any one of the following methods:—
    (a) Adhesive Tape.
    (b) MAGNETS. (No. 22)—these can be taped to the charge, or, if the charge is covered with rubberised fabric they can be stuck to it with Bostik, though additional fixing is recommended.
    (c) String.
    (d) Wedges.

The charge and firing system should be kept separate as long as possible. The firing system should always be duplicated.

11.—Double Initiation.
Firing systems should always be used in pairs. For example, there should always be two detonators, two pieces of safety fuse and two igniters. Similarly, Cordtex, whenever possible, should be doubled. This will avoid failures.

D. HOW TO ATTACK VARIOUS TARGETS.

The best results from explosives will only be obtained if:—

(1) Choice of target is right.
(2) The placing of the charge on the target is correct.
(3) The charge is prepared properly.
(4) A correct estimate is made of the amount of explosive that is required.

It is not possible to give detailed instructions in this small pamphlet, but sufficient knowledge of these four points to cause useful damage can be acquired quite quickly if expert advice is sought and the following instructions are carried out. These instructions describe very briefly how to place simple charges on the most valuable objectives.

1.—The Choice of Target.

The target selected must be one which, if damaged, will cause serious interruption for an appreciable time to the supplies being made for the enemy or to his lines of communication. Care must be taken to ensure that even temporary repairs will take some time to effect. Steel is very much easier to repair by welding than cast iron, which is shattered by the explosive. In using explosives on machinery, it is therefore best to attack cast iron parts.

How to recognise cast iron.
(i) It is usually rougher than steel plate.
(ii) It is usually thicker.
(iii) The maker's name may be cast in raised letters. With a part made of steel plate, the maker's name will usually be on a separate plate screwed on to the machine.

In a factory there are usually a few machines which are really vital. If you can, consult someone in the factory who knows which ones are really important. Most of the machine tools are not worth destroying. The power supply is usually worth attacking. Attack the transformers rather than the cables or overhead transmission lines. Large electric motors are good targets, but small ones are not. Locomotives are good targets because there is a great shortage on the Continent.

Road transport, and supplies such as ammunition, spare parts, tyres, food and clothing, petrol and other stores destined for use of the German Army are also vital targets, but some of these are better dealt with by fire.

2.—The Correct Position of the Charge.

(a) Machinery.
(i) If you can, destroy the machine when it is running, because more damage will be done. Place a charge on the support for one or more of the bearings. (Diagram 40). This support will nearly always be of cast iron.
(ii) If the machine is stationary, attack the largest casting, providing it is iron, and not steel. (Diagram 40).

(iii) **Machine Tools**—Destroy one of the main sliding surfaces or the gear box.

(iv) Cast iron pipes, pumps and other vessels are easier to destroy if they are filled with liquid. Steel structures of this type are more difficult, and you should not attack these without expert advice.

(v) If attacking large iron castings—(e.g. low pressure cylinders or large size turbines) it is better to use a number of small charges placed about ½ metre apart rather than one large one.

(vi) Electric motors and generators should be attacked, if possible, when running, by placing the charge below the bearing support. (Diagram 40). If they are stationary, the charges should be placed on windings of the rotating part, and pushed well into the machine. This may not be possible with alternating current machines, and in this case, the charge should be pushed through an opening in the frame, and placed on the iron core which will be seen inside. For large motors, use more than one charge placed about ½ metre apart.

(vii) **Locomotives**—Place the charge on the back end of the cylinder. Always place the charge on the right hand side cylinder; this will prevent the Germans using an undamaged cylinder from another locomotive to replace the damaged one.

(b) **Steel Work.**
(i) Most steel structures are made from steel sections. When cutting a section, make sure that the corner between the flanges and the web is well filled with explosive. (Diagram 41).

(ii) When attacking steel structures built into concrete foundations, such as supports for coal conveyor or an electric transformer line pylon, cut the main steel work as close to the concrete as possible but above any reinforcement there may be near the bottom. This will make it more difficult to repair. To ensure that the structure will fall over, a length of about 1½ metres should be cut out of the main steel supports on one side.

3.—**Targets and the size of the Charge Required.**
It is best to memorise the effect of each charge. If the objective is well known, the correct charge can then easily be selected. If the target is unknown, a small range of charges should be taken and the appropriate one selected.

The following notes are very brief and are intended only to act as a guide.

(a) **The 1/2 lb. Clam.**
It will:—
(i) Destroy a cast iron pump filled with liquid and which is not more than 1 metre in height.
(ii) Destroy a cast iron pipe, up to 60 cms. diameter if it is filled with liquid.
(iii) Destroy a locomotive cylinder.
(iv) Cut an angle iron.
(v) Destroy a small bearing pedestal, which is not more than about 15 cms. in width.
(iv) Destroy the petrol tank of a car or lorry and ignite the petrol. (A smaller charge of 4 ozs. of P.E. fitted with magnets will also do this).

(b) The General Purpose Charge, or half a 1 1/2 lb. Charge.

The General Purpose Charge is waterproof. Owing to its rigid form it is not so adaptable to targets with curved surfaces as a more flexible charge.

The charge should be considered, for effect, as the equivalent of half a 1½ lb. charge. It will destroy all targets enumerated in section (a) above

(c) The 1 1/2 lb. Charge (either pattern).

This charge will destroy:—

(i) Large cast iron pumps, cast iron pipes up to 1½ m. diameter if filled with liquid, bearing pedestals, of which the short side is not much longer than the length of the charge. (If the pedestal is larger than this, two charges should be used.)

(ii) An electric transformer, of which the long side is not greater than 1½ m. Place the charge in the centre of a long side and at least a quarter of the way up from the bottom of the tank.

(iii) The slides of a large machine tool.

(iv) A medium sized electric motor. If stationary, use two charges— about half a metre apart—if the motor is large.

(v) A steel joist 30 x 15 cms.

(vi) Small cast iron flywheels.

(d) The 3 lb. Charge.

This charge will destroy:—

(i) Heavy cast iron flywheel.

(ii) Electric transformer, of which the long side is between 1½ and 2 metres in length. Above these sizes, use two or three charges, one in the centre and the other(s) half way between the centre and the end of the tank.

(iii) A steel girder 40 x 15 cms.

(e) The Rail Charge.

The method of preparing this charge has already been described. It should be used with two FOG SIGNAL FUSES (No. 23) each of which has a spring snout (1) to hold a detonator (2.) The Cordtex tails (3) from the charge should be taped (4) to the detonators. (Diagram 42.)

The whole charge should be placed on the OUTSIDE of the rail, (Diagram 43) with the fog signals (1) on the side from which the train is coming. In the case of double tracks the charge should be placed on one of the inner rails, so that when the train derails, it blocks both tracks. (Diagram 44.)

The best place to stage a derailment is either a tunnel or a cutting. This makes the removal of the wreckage more difficult.

(f) The Made Up Rail Charge.

This charge, as already described in Section C, sub-section 7, is ready for immediate use. The fog signal (1) and the detonator (2) are packed separately, and a special spring snout attachment (3) is provided for fixing the Cordtex (4) to the detonator. Taping is unnecessary. (Diagram 45.)

The spring snout is simply pushed over the end of the detonator until the end of the detonator is hard up against the copper sealing cap (5) on the end of the Cordtex.

Special Note:

When placing fog signals on a railway line, the snouts and detonators must always point to the outside of the track. If they point inwards, the flange of the wheel will cut them off, and the charge will not fire.

E. DELAY ACTION.

The operator will normally wish to be clear of the scene of action when the incendiary bomb or explosive charge fires. For this purpose he should initiate these with a delay action device.

The device supplied is:—

1.—THE PENCIL TIME FUSE. (No. 24.) (Diagram 46.)

This is operated by the action of a liquid which corrodes a wire. When the wire is eaten through, a spring loaded striker is released, and is driven forward, by the spring on to a percussion cap. The pencil time fuse is shaped like a pencil; one end consists of a copper tube (1); on the other end is a spring steel connector (2.) The corrosion takes place in the copper tube, and the percussion cap is inside the spring steel connector. The firing of the cap will ignite a detonator (3) or piece of safety fuse which has been inserted into the connector.

Instructions for Use.

(i) See that the pencil time fuse is in order by looking through the inspection holes (4). It should be possible to see clearly right through both holes. If the view is obstructed, the Pencil must not be used.

(ii) Connect the pencil time fuse to the safety fuse or detonator, whichever is to be used, by simply pushing the end of the fuse (or detonator, open end first) as far as it will go into the spring steel connector. The connector will grip the safety fuse or detonator so that it will not fall out.

(iii) Link up the safety fuse or detonator to the incendiary bomb or explosive charge as already described.

(iv) Set the fuse in operation by crushing the copper tube carefully by hand. Squeeze with the fingers and thumbs of both hands but do not bend, or the wire may snap. (Diagrams 47, 48 and 49.)

(v) While all this is being done, the pencil time fuse cannot fire because the safety strip (5) is in position. This strip is below the inspection holes in the centre portion. It is withdrawn by straightening the end and pulling out.

(vi) The pencil time fuse will fire after a certain time from the moment of crushing. The period of delay for the time pencil fuse is indicated by the colour of the safety strip.

Red Safety Strip	½	hour.
White Safety Strip	2	hours.
Green Safety Strip	5½	hours.
Yellow Safety Strip	12	hours.
Blue Safety Strip	24	hours.

These timings are approximate, as they are for temperatures of 20°C.

Where the temperature is colder than this, the delays will be longer and where it is higher they will be shorter.

Do not leave the safety strip behind; it will tell the enemy the timing of the fuse.

F. INCENDIARISM.

When starting fires, either with home-made or manufactured incendiaries, the following general rules should be applied when possible:—

(1) Provide as much as possible of kindling material which will burn easily, e.g. thin pieces of dry wood, paper, curtain material, etc.

(2) Oily rags, cotton waste, bags, etc., form good "wicks" to prolong the flame.

(3) Good ventilation is essential. Open doors and windows (but do not interfere with blackout if this will reveal the presence of the fire more easily). In multi-storey buildings, start the fire near the base of stair wells or lift shafts, provided there is plenty of combustion material. The draught should be towards the most vital part of the target. Windy weather conditions will be favourable.

(4) Start the fire near vertical surfaces which burn more easily than horizontal ones.

The following special stores, which may be used either with a time delay or for immediate use, are available in containers for starting fires:—

1.—THE FIREPOT. MK. IA. (No. 25.) (Diagram 50.)

This is a magnesium incendiary bomb filled with Thermit. On ignition vigorous burning with showers of sparks lasting about 8 seconds is followed by slow burning for 10-15 minutes.

The Incendiary is protected from damp by a tin lid (1) held in place by tape (2).

To prepare the bomb for use, remove the tape (2) which holds the lid in place and lift off the lid (1). (Diagram 51.)

In the centre of the top of the bomb is a match head (3) and 2 short lengths of safety fuse are coiled round the side (4). For immediate use, remove the striker board (5) from the underside of the bomb by peeling off tape (6) and strike across the match head (3). (Diagram 52.)

To give a short delay, attach copper tube igniters to the ends of the safety fuse cutting off 1 cm. before they are fitted on. For a longer delay attach a Pencil time fuse to each piece of safety fuse.

The Firepot is a good general purpose incendiary which will readily ignite anything inflammable, but it has not the penetrating power of the 2½ lb. Thermit described below.

The Mk. I Firepot is exactly the same as the Mk. IA but has no lid and no match head.

2.—THE 2 1/2 lb. THERMIT. (No. 26.) (Diagram 53.)

Contained in a black metal tin is quick burning thermit. Remove the lid by peeling off the tape.

The bomb may be fired by means of the copper tube igniter attached to the Bickford fuse or these may be removed and time pencils attached in their place.

A longer delay may be used by removing the cap and Bickford fuse and inserting the required length of Bickford into the igniting material container.

The bomb burns quickly, generating intense local heat and fierce flames. It will penetrate mild steel plate 3.1 mm. thick.

Its main uses are:—

(1) For igniting petrol and other inflammable oils in closed containers, e.g. army petrol cans in stacks.

(ii) For attacking difficult non-inflammable targets of compressed material, in conjunction with other incendiaries.

(iii) For demolition work, e.g. for attacking telecommunication cable, electric motors, etc.

3.—THE INCENDIARY BLOCK. (No. 27.) (Diagram 54.)

This incendiary consists of a transparent case (1), which is watertight, containing the incendiary substance which is in two parts (2). The lower half consists of a substance which produces oxygen. It is white in colour and must always be placed in contact with the target. To enable this to be done in the dark, the top of the case has a single raised rib (3). This rib must always be uppermost when the incendiary is placed.

To ignite the incendiary, tear off the strip (4) covering the double safety fuse tails, open the fuse compartment, and bring the tails (5) out into the open. (Diagram 55.) Each is fitted with a copper tube igniter which can be lit with a striker board.

The copper tube igniter can be cut off and Pencil Time Fuse substituted if desired.

The Incendiary is particularly suitable for burning through wood, e.g. the lid of a packing case. The initial burning quickly produces a hole and the substance in the top half of the block then becomes molten and burning and passes through the hole, setting the contents on fire. It will not penetrate wood more than 18 mm. thick.

It is a handy general purpose incendiary but it must be carefully placed on suitable combustible material. Good targets are floors or ceilings of rooms, barns, storage sheds, warehouses—concealed in the spaces between stacked packing cases—timber yards, railway wagons and haystacks.

It will also burn in confined spaces.

For use against fuel oil, the incendiary should be wrapped in sacking and if it is likely to become swamped by a flow of oil, it should be lashed to a wooden raft which will float on the oil.

4.—THE POCKET TIME INCENDIARY. (No. 28.) (Diagram 56.)

This is a delay action incendiary.

The incendiary material is contained in the two outer cylinders of the incendiary. In the centre is a mechanism similar to that of the Pencil Time Fuse.

To operate the device:—

(a) Peel off the protective cellophane which covers the slot (1) in the front of the case. In the slot is a coloured wooden stick (2) which acts as a safety pin. If this stick is jammed, the mechanism is faulty and the device must not be used. If the mechanism is sound, the stick should fall out quite easily. (Diagram 57.)

(b) Insert a coin (3) or the back of a clasp knife into the slot and press firmly down until the glass ampoule inside the copper tube is heard to crush. (Diagram 58.)

The device is then armed.

The length of delay is indicated by the colour of the safety stick and the range of delays is the same as for the Pencil Time Fuse

The Pocket Time Incendiary burns fiercely for about 50 seconds. It must therefore be placed in contact with combustible material. If it is strapped to a bottle of petrol or paraffin, it will crack the bottle without noise and start an appreciable fire which will burn for some time. (Diagram 59.)

5.—Linking Incendiaries.

A special fast burning fuse called INSTANTANEOUS FUSE (No. 29) is provided for this purpose. Like safety fuse, it has a black filling but it can easily be distinguished by its orange covering. It must always be initiated by Time Pencil, Pull Switch, or Safety Fuse, not by match, fusee or copper tube igniter direct.

The method of linking firepots and 2½ lb. Thermit Bombs is:—

(i) Cut a suitable length of instantaneous fuse.

(ii) Space out the incendiaries as required along this length.

(iii) Place the instantaneous fuse across the top of each incendiary and cut a notch (1) in the fuse where it crosses the safety fuse tails of the incendiaries, exposing 5 mm. of the black filling. In the case of the firepot cut a third notch over the match head. (Diagrams 60 and 61.)

(iv) Where the instantaneous fuse crosses the safety fuse, expose about 5 mms. of the black powder filling in the safety fuse (2), by paring away the covering from one side carefully with a knife.

(v) Tape the instantaneous fuse and the safety fuse together so that the exposed fillings are in contact.

(vi) Tape the instantaneous fuse down over the match head in the case of the firepot.

(vii) Repeat this process until all incendiaries are linked together on to the same piece of instantaneous fuse.

6. THE LIMPET. (No. 30.)

This is a charge designed for attacking ships or barges from the outside and below the waterline.

1. The Limpet. (Diagram 62.)

The charge is enclosed in an oblong metal box (1) with a set of magnets (2) fitted to each side. There is a screw collar (3) at each end to receive the fuses. If received empty, unscrew the large metal cap (4) at the end of the box, and fill with explosive. The best fillings are Plastic or 808 (the latter should be kneaded into the container but the Burster will not fire it at high speed and a primer must therefore be added). On the top of the case is a slot (5) to take the end of the placing rod. The holes in the ends of the limpet are fitted with wooden rectifiers. Their use is described in Sub-Section 5.

The charge is fired by means of a delay action fuse known as the A.C. Delay Fuse, which fires a burster.

2. THE A.C. DELAY. (No. 31.)

This delay action underwater fuse is supplied in a tin box. There are three parts:—

(a) **The Fuse Body** (1) which contains a striker and spring, and has a detachable top (2). A safety pin (3) passes through the screw plunger (4).

(b) **The Burster** (5) with percussion cap (6) attached. (Diagram 64.)

(c) **The Ampoules** (7)—six are provided, each filled with a clear liquid of different colour.

To assemble the fuse:—
(a) Unscrew the detachable top (2).
(b) Insert one ampoule (7) point downwards.
(c) Replace the detachable top (2).
(d) Screw on the Burster (5).

The drill for these operations is described more fully in section 5.

The time of the delay is indicated by the colour of the liquid in the ampoule. This delay varies according to the temperature of the water. At 15°C. the approximate timings are:—

Red	4½ hours
Orange	7½ hours
Yellow	15 hours
Green	26 hours
Blue	42 hours
Violet	5½ days.

A full table giving the delay at various temperatures is included in the box.

3.—Instructions for the use of the Limpet.

The limpets are designed for attacking ships or steel barges. They will adhere to the metal plate of a ship, and should be placed more than one metre below the water line; to do this, the placing rod described below is of great value.

The limpet will adhere to the side of a clean ship (i.e. one not covered heavily with weeds or barnacles) and will remain in contact, in a calm sea, even if the ship is moving at a speed of 10 knots. On a dirty ship, or when there is a heavy swell, too much reliance should not be placed on this adhesion. Some barnacles do not thrive in deep water, so may sometimes be avoided if the limpet is placed at a greater depth.

In selecting positions for the placing of limpets, the after holds should be considered first, as a ship will sink faster when these holds are flooded. It depends, however, on what cargo the ship carries. If it is cotton bales or other bulky materials, the likelihood of flooding the hold is very remote. On the other hand, iron ore or other similar mineral will only occupy a comparatively small space as compared with the bulky materials of the same weight, and hence allow the holds to flood freely. (Diagrams 65, 66, 67.)

The engine room is a good position, provided the ship is not fitted with wing bunkers. Large merchantmen and ocean-going vessels have watertight bulkheads between the holds, and therefore to sink such a ship, more than one limpet is required.

As a guide to the operator, for ships over 3,000 tons, attack both after holds and the engine room. For larger vessels, the forward holds and the other side of the ship should be attacked. The engine room is usually below or near the funnel.

4.—Drill for linking Limpets.

When placing more than one limpet on a ship they must be linked together with detonating fuse, otherwise the explosion of the first one may dislodge the others. The method of linking limpets is as follows:—

A.—Two Limpets.

Before setting out:—

(a) Unscrew end covers (1) and remove wooden rectifiers (2) from them. Scoop out approximately 5 cms. of P.E. from these ends.

(b) Thread corks (3) with double holes over the required length of double Cordtex. Push the Cordtex through the holes in the end covers and thread primers (4) on one of the Cordtex leads, at both ends. Knot the ends of the Cordtex. (Diagram 68.)

(c) Embed the primers in the explosive of each limpet and pack around the primer with the explosive originally scooped out until the P.E. is almost flush with the top.

(d) Grease the threads and screw in the end caps, using the issued SPANNER (No. 32) to make the final tightening, then grease over the outside union of each cap.

(e) Push corks into holes. Tape them well to the small projections and finally grease over the tape. (Diagram 69.)

B.—When more than two limpets are used in series, to be fired simultaneously.

The intermediate limpets will require to have the rectifiers and end cap removed, and all the explosive scooped out so that the double Cordtex can be threaded all the way through, with a primer at each end.

In the intermediate limpets, two corks will be required.

5.—Drill for Preparing and Fusing Limpets.

(1) Before setting out on an operation:—

It is extremely important when limpets are used that a set drill for preparing and fusing them should be carried out:—

(a) Fill the limpet with P.E. or 808 (if not already filled).

(b) Plug a wooden rectifier into each end of the limpet.

(c) Withdraw the rectifiers, reinsert and withdraw once more gently to ensure that the holes in the charge are intact and free to take the burster.

(d) Take one A.C. Delay. Unscrew the removable top and check:—
    (i) That the striker and spring are in the compressed position.
    (ii) That all screw threads are undamaged.
    (iii) That the safety pin is not rusted and can be removed easily.

(e) Take one burster. Check the cap visually and screw the burster on to the Delay tightly.

(f) Then, but not before, grease thoroughly all the exposed threads with LUTING (No. 33). This applies only to the 28a Burster. Special instructions for the Type 6 Burster will be found at the end of Section G.

(g) Grease the screw threads which take the removable top, ensuring that no excessive grease falls into the body of the delay or exudes over the edge. It is best not to bring the grease right to the top of the threads.

(h) Take one ampoule, giving the required delay, and insert it into the body of the delay WITH THE POINTED END DOWNWARDS, and resting on the cotton pads.

(i) Screw on the removable top.

(j) Insert the Burster gently into the hole formed by the rectifier at one end of the limpet. Engage the screw threads, screw home and tighten with a spanner.

(k) Grease thoroughly the external union of the Burster with the limpet.

(l) Remove cap once more to check finally that the ampoule is sound and is situated the correct way up. Replace the top and grease the junction. The limpet is now ready for use.

(Note.—Repeat this drill for all A.C. Delays until all the limpets are fused.)

(2) On the way to the Target.

At a convenient point during the approach to the target, taking into account the length of delay being used and the time the approach is liable to take, the fuses on the limpets should be set in operation as follows:—

(a) Remove safety pins, drop into the water.

(b) Screw down caps of each A.C. Delay until the ampoule is heard and felt to be crushed THEN SCREW ON TO THE FULL EXTENT OF THE THREAD,, even if cracking has been heard.

6.—THE PLACING ROD. (No.34.)

This is a collapsible rod with "shoe" which can be slipped into the slit on top of the limpet; the other end of the rod is hooked. When fully extended the rod measures 1 m. 27·5 cms. (Diagram 70.) It is constructed to assist in placing the limpet on to a ship from a boat lying alongside.

The limpet can be attached to the "shoe" and lowered into the water. By gently pressing the rod towards the ship, the limpet can be brought into contact with the plates. When it adheres, with a slight downward pressure the shoe can be freed and the rod withdrawn. (Diagram 71.)

7.—THE MAGNETIC HOLDFAST. (No. 35.)

This is provided to give a temporary hold to the boat while the placing of the limpet is carried out.

8.—Type 6 Burster.

There are two models, the Mark I which is used with the normal A.C. Delay and the Mark II which is supplied with a special A.C. Delay fitted with rubber washers.

With the aid of a coin, unscrew the protector plug (1) from the end of the Burster. Then screw the Burster on to the A.C. Delay. If the A.C. Delay has no washers proceed as for the 28a Burster. If there are washers, luting should on no account be used.

The Burster should be screwed on hand tight only in each case.

H. SPECIAL MECHANISMS.

These mechanisms are so constructed that when the spring loaded striker is released, it drives forward and fires a percussion cap contained in the snout.

This snout can be unscrewed and the mechanism can be operated for testing. When doing this, point the snout end towards the floor as the striker will fly out with considerable force. It can then be reset by forcing back the striker into position with a pencil or large nail.

The cap will fire a detonator or Bickford safety fuse inserted into the spring snout.

Each device has a safety pin.

It is a sound principle to use these devices in pairs.

1.—THE PULL SWITCH. (No. 10.) (Diagrams 15 and 72.)

As already explained, this mechanism is an excellent means of igniting safety fuse. It is considered here, used in conjunction with a trip wire, as a Booby Trap mechanism, where a detonator is inserted direct into the snout.

Instructions for Handling.

Attach the mechanism, by means of a length of string or wire passed through the two holes of the mounting strip, to some firm object.

Allow it to hang loose so that a direct pull can be exerted on the tail.

Connect the .014 TRIP WIRE (No. 36) to the tail of the switch and stretch it across the path along which you anticipate the enemy will come. Never have the wire too tight.

Place or bury your charge in a position where it is likely to do the maximum amount of damage

Then lead from the charge a length of cordtex and tape this to a short length of cordtex attached to the detonator with an overlap of about 5 cms.

Withdraw the safety pin and the trap is set.

2.—THE PRESSURE SWITCH. (No. 37.) (Diagram 73.)

This store is useful as a booby trap placed under objects which the enemy is likely to walk or drive across.

For derailing trains as a substitute for the Fog Signal it has the advantage that it can be concealed and it should be used with a larger charge buried in the ballast or in a tunnel. On the other hand it is not so certain in operation and should only be used if absolutely necessary.

(a) As a Booby Trap.

A charge should be prepared and buried or otherwise placed with a length of cordtex leading to the point at which the pressure switch is to be placed.

The pressure switch should then be placed in a position where it will be operated when a pressure of 18 kilos. is exerted on it, e.g. under doormats, in a path or road, buried level with the surface.

Where two pressure switches are used they should be placed some distance apart, with a board or piece of metal laid across them, in such a way that if pressure is applied to any part of the board, one or both switches will fire.

This board must be light, otherwise it will fire the switch. Both of the switches should, of course, be connected to the same charge.

(b) In Train Derailment. (Diagram 74.)

Bury the mechanism in the ballast of the track, resting it on a firm base. Adjust the extension rod, screwed into the top of the lid, until the point comes in contact with the bottom of the rail.

Care must be taken that the point of the rod is touching the line but not pressing too firmly, or else the charge may be detonated prematurely. Care must also be taken to ensure that the rod is placed in an upright position or else the pressure will not be applied correctly.

Before concealing the switch, do not forget to remove the safety pin.

3.—THE RELEASE SWITCH. (No. 38.) (Diagram 75.)

This mechanism is designed to operate when a weight is removed from the tail.

The weight placed on the tail should exceed 1½ kilos, and if it is insufficient the safety pin will not come out easily.

Ensure that the weight is correct before connecting up your cordtex from the charge to the short length attached to the detonator, which is inserted into the spring snout.

Place the switch underneath anything which is likely to be lifted up by the enemy.

Do not forget to remove the safety pin.

NOTE.—A short length of cordtex should always be attached to the detonator on the mechanism before it is placed in position. This short length can then be connected to the length from the charge. This prevents the mechanism being accidentally disturbed by the operator after it is set.

(See point A in each diagram.)

I. MISCELLANEOUS ITEMS.

1.—THE No. 82 GAMMON TYPE GRENADE. (No. 39.)

This is a grenade which can be made up by the user when he requires it. It explodes on impact. For use against armoured cars, lorries and light tanks it should be filled with Plastic, but it can be filled with any suitable explosive and used as an anti-personnel weapon.

The grenade is made up of three parts:—
 (a) **THE PERCUSSION FUSE AND CLOTH BAG.** (No. 39.) (Diagram 76.)
 (b) **THE PRIMER CUP.** (No. 40.) (Diagram 77.)
 (c) **SPECIAL SHORT No. 8 DETONATOR.** (N.B.—The Ordinary detonator will NOT do.) (No. 41.) (Diagram 78.)

Instructions for Assembling.
 (i) Mould approximately 1 kilo of Plastic into the shape illustrated leaving a hole for the primer cup. (Diagram 79.)
 (ii) Insert a detonator into the hole in the centre of the primer—closed end first. Always use the special detonator.
 (iii) Screw the primer cup on to the Fuse.
 (iv) Insert the primer cup into the hole in the explosive and draw the cloth bag over the explosive. (Diagram 80.)

Preparing the Grenade for Throwing.
 (i) Unscrew the black bakelite cap of the fuse, taking care that the white tape does not unwind.
 (ii) Hold the grenade in the right hand with the explosive resting on the palm of the hand and the fuse held with the thumb and forefinger pressing firmly on the tape. (Diagram 81.)

Instructions for Throwing.
Throw with an action similar to throwing a ball. The tape will unwind in flight, pulling out the safety pin and the grenade will explode on impact. (Diagram 82.)

Instructions for Use.
The grenade is best used for attacking armoured cars, staff cars and lorries. It may also be used against light tanks, but it should not be used against heavy tanks.

Accurate throwing is not possible over 20 metres owing to the weight of the grenade. Therefore use the grenade at close range.

Safety Precautions.
If the tape and safety pin should accidentally come out before the grenade is thrown, no attempt should be made to throw the grenade as there is a danger that it will go off in the hand. The safety pin and tape should be replaced before the grenade is thrown.

2.—The General Purpose Charge used as a Grenade.

The PERCUSSION FUSE (ALLWAYS) (No. 42) for the General Purpose Charge is the same as that of the No. 82 Gammon described above. An ordinary detonator should be inserted, closed end first, into the hole at the end of the charge with the threaded socket. The fuse should then be screwed into position.

The Grenade should be thrown in the same way as the No. 82 Gammon.

3.—THE TYRE BURSTER. (No. 43.)

The Tyre Burster is a small bomb. It consists of a small but powerful charge enclosed inside two interlocking circular metal cups. When pressure is exerted on the device, these cups telescope, and the mechanism operates to fire a detonator, which sets off the charge. Note that the top and bottom of the tyre burster are camouflaged with tape.

It is particularly suitable for attacking pneumatic tyres and the explosion is sufficient to puncture the heaviest tyre.

Preparations for Use.

(a) Take a burster out of the box in which it is packed.

(b) Lift the square piece of tape marked with a white spot, revealing the lead plug. (Diagram 83.)

(c) Unscrew the lead plug, using a screw-driver or small coin.

(d) Examine the face of the lead plug. If a small hole has been punched in it, the Tyre burster is no use and no attempt should be made to use it. (Diagram 84.) Even if the lead plug is unmarked, always make sure that the striker is not projecting into the threaded detonator pocket. The Tyre Burster in Diagram 85 is sound.

(e) If you are satisfied that the Tyre Burster is in order, take a DETONATOR (No. 44) from the tin, and screw it into the detonator pocket, tightening it up with a screw-driver or small coin. (Diagram 86.)

(f) After the detonator has been inserted, replace the square piece of tape marked with a white spot.

(g) The Tyre Burster is now live and ready for use.

(h) Do not remove the tape which holds the metal cups together, since it is designed to keep the Tyre Burster waterproof.

Instructions for Use.

A pressure of 45 kilos is required to set off the burster.

(i) For Use against Vehicles on Roads.

(a) On metalled roads use a number of bursters, arranged at least 1½ metres apart, as far as possible in echelon formation. They should not be closer together than 1½ metres, or the blast from one will displace the others. The best way to conceal them is to put them into holes in the road or into puddles. Otherwise camouflage with dust or dirt.

(b) On loose cinder or sandy roads, bury the bursters just below the surface. Place a stone or other hard object under each burster, to prevent it being pressed by the vehicle into the soft ground without exploding.

(c) Always put the bursters where they are likely to do damage to as large a number of vehicles as possible.

(ii) **Other Uses.**
   (a) Under the tyres of vehicles parked in a field—the grass will provide excellent camouflage.
   (b) Under the tyres of aircraft dispersed on airfields.
   (c) On the driver's seat of a car placed under the covering.
   (d) Under a carpet or mat.

(iii) **How to make up a Home-made Anti-Tank Mine.**
   There are two ways of doing this:—

   Method I. (Diagram 87.)
   Construct a wooden box (1) 20 cms. × 20 cms. × 15 cms. to hold about 4½ kilos of explosive.
   Make a lid (2) which overlaps the sides. Soak the box and lid in creosote.
   Drive nails (3) through the lid at intervals and bore suitable holes to take them in the wooden edge of the box.
   Inside the box in the bottom, hollow out a recess to take a Tyre Burster (4).
   Arm a Tyre Burster and place it in the hollow.
   Cut a length of broomstick (5) which when rested on the Tyre Burster will project slightly above the level of the top edge of the box.
   In the lid and centred over the hollow in the bottom of the box, fix a cup (6) to take the top of the broomstick (5). Make this out of a shaving stick tin or other suitable metal cylinder.
   Resting the broomstick on the tyre burster, pack the box with explosive (7) level with the lid.
   Put on the lid of the box so that the broomstick enters the cup (6) and press gently down until resistance is felt.
   Bind the box with wire (8).

   Method II. (Diagram 88.)
   Dig a shallow hole 25 cms. square and 10 cms. deep.
   Place two lengths of wood (1) 7 cms. wide down opposite sides of the hole.
   Place four armed tyre bursters (2) on the lengths of wood one at each corner of the hole.
   Place explosive (3) into the hole and round the bursters, bringing it to a level just above the top of the bursters.
   Place a wooden board (4) 25 cms. square over the explosive and cover with 3 cms. of earth.
   If these mines are laid in places where Allied Forces may pass, some metal object should be buried near the mine, and this will help in its detection.

4.—THE 36 OR MILLS GRENADE. (No.45.)

The grenade is filled with explosive which, on detonation causes it to break into a number of small pieces all of which can kill up to a distance of some 20 metres, and may even wound at considerably greater distances. It can be thrown up to 30 metres, but the thrower must be able to get behind cover before the grenade explodes. After the striker is released, there is a four second delay before detonation.

(1) Description. (Diagrams 89 and 90.)
The grenade consists of an iron body (1) filled with explosive (2). Inside in the centre sleeve is a spring loaded striker (3) which is held up by a lever (4) and safetypin (5) which passes through holes in the shoulder (6). At the base of the grenade is a base plug (7) which can be unscrewed.

THE DETONATOR. (No. 46) (8) with cap (9) are always packed separately, and should only be inserted when the grenade is being prepared for action. The grenade should never be stored with the detonator in place.

(2) Stripping and Cleaning.
 (i) Remove the base plug (7) using the tool provided.
 (ii) Ensure that the cap and dentonator are not in the grenade.
 (iii) Hold the base of the grenade against the body; grasp the lever (4) firmly, and take out the safety pin (5).
 (iv) Release the lever (4) gently and remove the striker and spring (3).
 (v) Clean all parts thoroughly to remove the grease.

(3) Reassembling.
 (i) Replace striker and spring in the centre sleeve and with a stout nail or metal rod, push them up until the slotted end (10) emerges through the hole in the top of the grenade.
 (ii) Slip the lever (4) into the slot, press it down into its recess on the side of the grenade and push in the safety pin (5).

(4) How to Arm the Grenade.
 (i) Take one cap and detonator and insert the detonator into the small hole next to the centre sleeve (11).
 Push it in as far as it will go, at the same time forcing the cap (9) into the open end of the centre sleeve.
 (ii) Screw on the base plug (7), making sure that it is firmly in position.

(5) How the Grenade Works.
The grenade explodes when, after removing the safety pin, the lever flies up, releasing the striker, which is forced down by the spring on to the head of the cap—this lights the fuse which, after four seconds, explodes the detonator and through this the rest of the explosive.

(6) How to Throw. (Diagram 91.)
 (i) Hold grenade in the palm of the hand so that the safety lever is held down firmly by your fingers.
 (ii) Still holding down the lever, remove the pin. SO LONG AS THE LEVER IS HELD, THE GRENADE IS SAFE.
 (iii) Draw the hand back and throw—making sure to throw high so as to avoid hitting obstacles. The lever will fly off as the grenade leaves your hand, thus causing the striker to fire the cap which lights the fuse.
(Note: Providing the lever has not been released, you may replace the safety pin and the grenade is once more safe.)

5.—ABRASIVE PASTE. (No. 47.)
This paste is provided for attacking bearings of machinery of all types. It is particularly useful for attacking axle boxes on Railway wagons.

Axle boxes are of many types, and it is impossible within the scope of this pamphlet to lay down any procedure for dealing with them which will be common to all.

The boxes must be opened, and for this an adjustable spanner will be required to remove the bolts which hold the cover in place. The abrasive paste should then be spread where it will come in contact with the axle and be drawn round in the bearing. It is advisable to obtain expert advice before this is attempted.

It is no good squeezing the paste into the filler through which the oil is inserted into the box, as it will merely sink to the bottom of the oil and will not reach the vital bearings.

6.—PLIERS. (No. 48.)

7.—VASELINE. (No. 49.)
This is suitable for use as a waterproofing material; or for mixing with Plastic when it is hard, to soften it.

8.—FIELD DRESSINGS. (No. 50.)

J. STEN SUB-MACHINE GUN (MARK II.) (No. 51.) 9 mm.

1.—General.
The Sten Gun is a very deadly close-range weapon but while it can be used effectively at longer range, it should not be regarded as a substitute for a rifle. Nothing has been wasted on exterior finish. The barrel and working parts, however, are well made, of good material, and can be relied on to fulfil the purpose of the weapon.

It can be fired at full automatic or at single shot, at user's option.

2.—Ammunition.
The Sten Gun fires the 9 mm. Luger (Parabellum) cartridge. Ammunition of this partcular type may be encountered of American, Belgian, British and German manufacture.

Four loaded magazines are provided with the gun.

3.—Assembling. (Diagrams I and II.)
For convenience, the gun is packed with its four main components separate These are:—
(A) Barrel, complete with cooling cylinder (1).
(B) Body.
(O) Butt.
(D) Magazine.

(See Diagrams I and II.)

To assemble A. B and C:—

(a) See that the magazine housing (a) is in the position shown in the diagram, i.e. horizontal and at right angles with the body of the gun. Grasping cooling cylinder (1), screw the barrel into the forward end of the body. If there is any difficulty in screwing the barrel fully home, lift up cooling cylinder catch (3). Screw in firmly but not so tightly as to make it difficult to remove again when necessary.

(b) Holding the butt (C) with the perforated triangular metal piece (4) downwards, engage the butt stud (5) in the groove (6) immediately above the back of the trigger-guard (7). Depress return-spring housing (8) and push butt upwards until the stud is as far in the groove as it will go. Allow return spring housing to rise.

4.—Preparation for Firing.

Pull back the cocking-handle (9) until it is opposite the safety-notch (10). Turn it upwards and to the left until it engages in the safety-notch (10). The gun is now in the " safe " position.

Decide whether you want full automatic fire or single shot. For full automatic press in button marked A, on right hand side of gun (11). For single shot, press in button marked R on left-hand side of gun. Single-shot fire is recommended since it secures greater accuracy and economy of ammunition. Though the trigger must be pulled for each shot, great speed of fire can be obtained with practice. If the gun is set for full automatic fire, it must be remembered that a continued pressure of the trigger will result in the discharge of the entire contents of the magazine. It is possible, of course, to control the full automatic action by alternate quick pressure and release of the trigger but this requires experience.

With the left hand, insert a filled magazine in the magazine-housing (2), taking care that the bullets point to the forward end of the gun. Push the magazine in smartly until it locks in position. Make sure that it is so locked by pulling outward on it. The magazines supplied with the gun are already filled. To re-fill them, see instructions (and diagram) for use of Magazine Filler.

With the left hand, disengage the cocking-handle from the safety-slot (9) and let it go forward, under control, until it stops in the firing position. The gun is now ready to fire.

In firing, the left hand must grip the cooling cylinder (1), thumb and fore-finger bearing hard against the front end of it. The fingers must be kept clear of the ejection opening (12). Never hold the gun by the magazine. The right hand grasps the butt in the same way as for a rifle.

After firing, engage the cocking-handle (9) in the safety-notch (10). Remove magazine by depressing magazine-catch (13) and withdrawing magazine. See that the breech is clear and close the gun.

If firing at a target, the tip of the fore-sight (14) should be focussed on the lower edge of the bull's eye. The eye automatically centres the fore-sight in the aperture of the back-sight (15). If firing at a human target, aim about the waist-line, if crcumstances permit. (See diagram III.)

6.—Stripping.

Remove magazine (D) by depressing the magazine catch (13) and pulling out the magazine.

Remove butt (C) by pressing in the return spring housing (8) and sliding butt downwards until clear of the grooves in the receiver.

Remove breech block (16) by pressing forward the return spring cap (8) about 1 cm. and turning it until it is freed from the receiver. Remove return-spring housing cap and return spring. Press trigger and pull back breech block (16), then pulling out cocking handle when opposite safety slot. Hold gun muzzle uppermost and breech block will drop out.

To remove barrel, pull cooling cylinder catch (3) up, at the same time turn magazine housing (2) downwards. Unscrew cooling cylinder (1) and remove with barrel. Re-assemble the gun in the reverse order.

7.—Stoppages and their Remedies.

(a) **Fault in feed.** Put cocking handle at safe. Remove magazine. Re-seat top cartridge in magazine, i.e. with its base against the rear of the magazine, or, if the top cartridge has come out of the magazine, discard it. Replace magazine and continue firing.

(b) **Mal-ejection.** Put cocking handle at safe. Remove magazine. Remove empty case. See that there is no live round in the breech. If there is, shake it out or replace magazine and continue firing.

(c) **Miss-fire.** Cock the gun, ejecting the faulty round in the process. Continue firing

(d) **Burst case.** Put cocking handle at safe. Remove magazine. Shake out fragments of the case Make sure that a bullet is not lodged in the barrel. Replace magazine and continue firing. If, for any other reason, the gun refuses to function, examine breech block and interior of receiver for dirt or extraneous matter and, finally, look for broken parts.

8.—Care and Cleaning.

(a) Before firing, remove all oil or grease from receiver breech-block and interior of barrel.

(b) After firing, remove barrel and clean from the breech end with a cleaning rod or improvised pull-through. Clean breech block, receiver, magazine platform and wherever else fouling has collected. Oil the barrel and wipe over all other parts with an oily rag.

9.—Storage.

If the gun is to be put away for any length of time without attention, grease liberally inside and out (vaseline or motor grease will do), wrap in grease-proof paper and store in a damp proof container, but do not forget to clean again thoroughly before use.

Magazines should be stored empty.

10.—Magazine Filler. (Diagram IV.)

To fill the magazine, fit the filler to the magazine so that the charger catch (f) engages in the magazine recess or hole (g). Place the four fingers of the left hand on the lever (a) so that the third finger goes through the hole (b) in the lever and the first rests on the toe (c). Depress the tail (d) on which the fourth finger rests, and insert a round, base first, in the opening (e) with the right hand. Raise the lever by means of the third finger and depress again to insert another round. Repeat this procedure until 28 rounds have been inserted If filling by hand, remember that, when in position on the gun, the slots on the magazine (n) face the butt of the gun. Ensure that ammunition is clean before loading.

K. PISTOLS. (No. 52.)

The pistols supplied may be of many different makes, but will be of two main types:—
  (1) Automatics.
  (2) Revolvers.

1.—Automatic Pistols.

These weapons automatically eject the spent cartridge after a shot has been fired and feed another round into the breech.

The rounds, which are rimless, are contained in a removable magazine, inside the pistol butt.

When firing, grip the pistol very tight.

Care and Cleaning.

In order to ensure that an automatic will be ready for use, and in order that it shall not fail the firer at a critical moment, it is essential that it should be kept thoroughly clean and free from rust, and dirt.

If it is stored away for any length of time it should be thoroughly cleaned and all the working parts should be oiled. Access to all parts, which cannot readily be stripped, should be sealed with grease.

The weapon should then be wrapped in grease-proof paper or oily rags and stored in as dry a place as possible.

The only stripping necessary is the removal of the barrel, the slide and the magazine. The trigger should never be stripped.

2.—Revolvers.

These weapons are semi-automatic only, and the rounds, usually six in number, are rimmed and are held in a revolving cylinder.

The action of squeezing the trigger, fires the round and revolves the cylinder, bringing the next round into the firing position.

After all the rounds have been fired, the cartridge cases remain in the cylinder.

The methods of opening revolvers differs considerably. In the case of the .38 Webley, this is done by holding the weapon in both hands, pressing down on a catch on the left hand side of the butt with the thumb and "breaking" the revolver. If the cylinder is loaded with spent cartridge cases, these will fly out and should be caught in the hand.

Care, Cleaning and Storage.

The stripping required is very simple. It is only necessary to take out the cylinder, which is usually released by removing a screw.

The same rules of cleaning and storage apply.

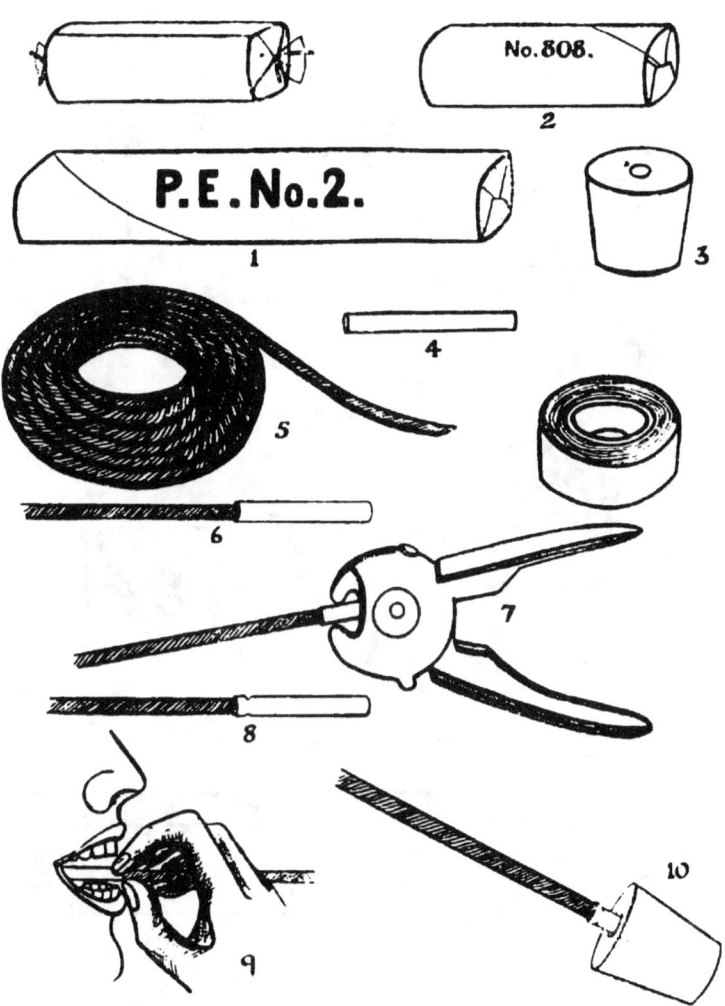

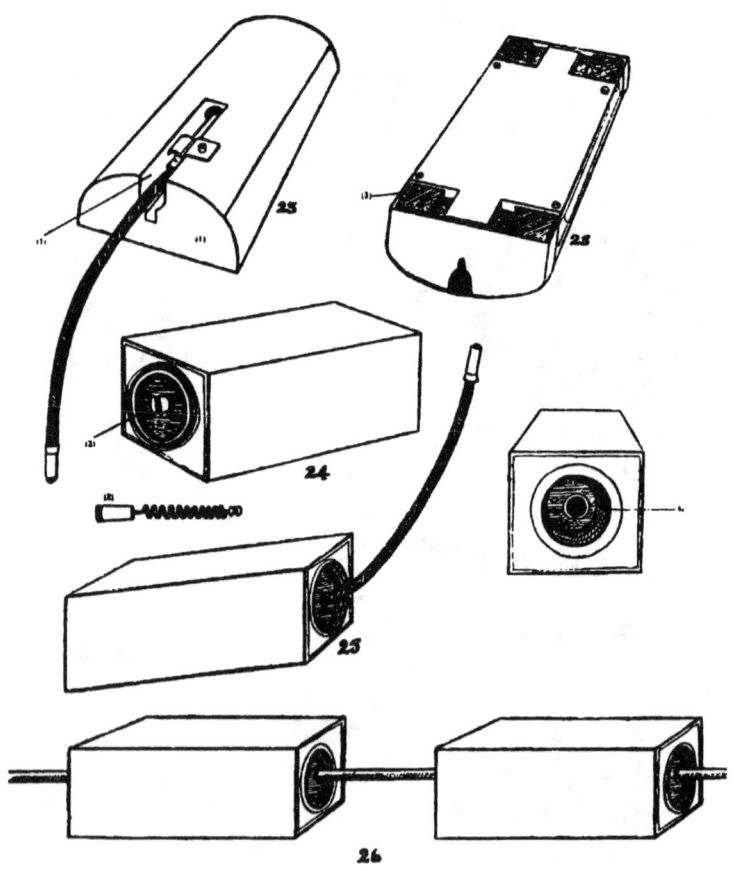

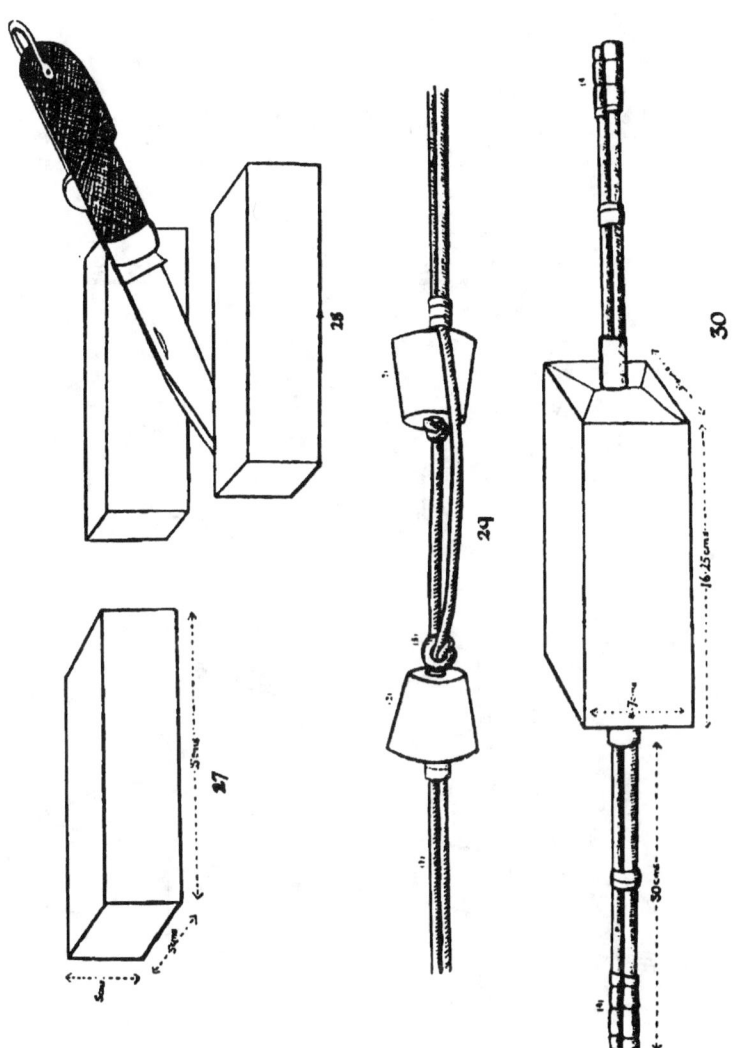

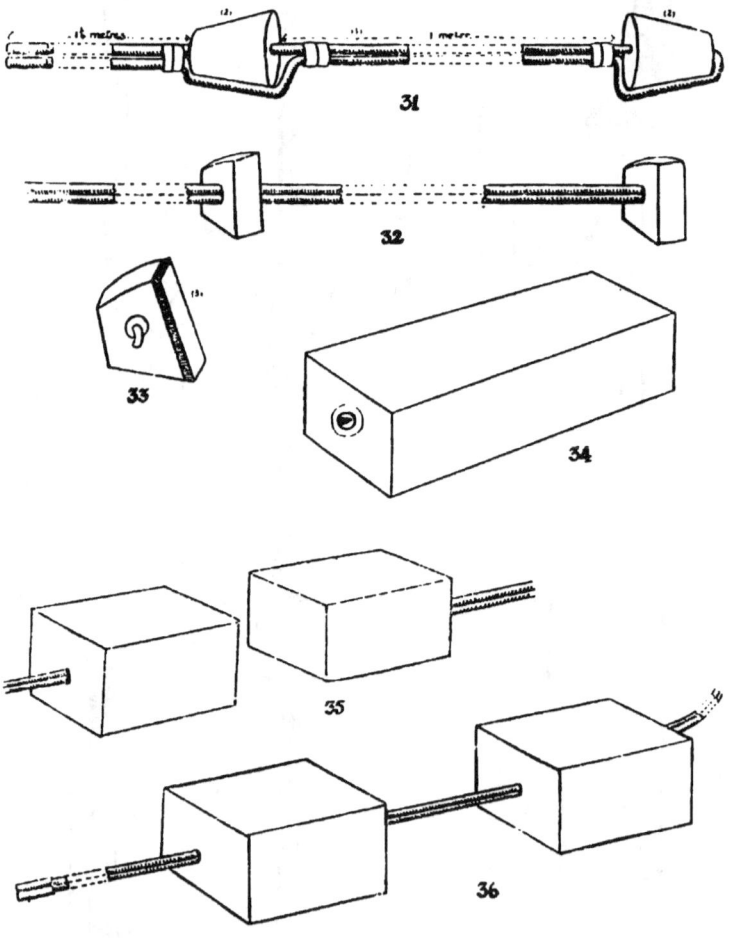

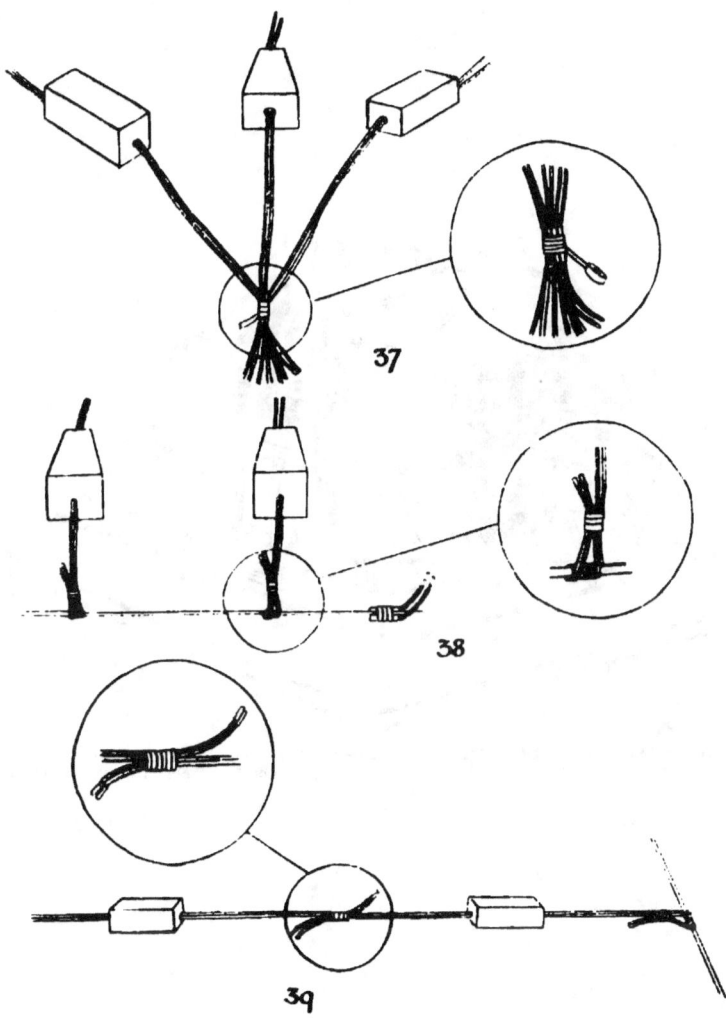

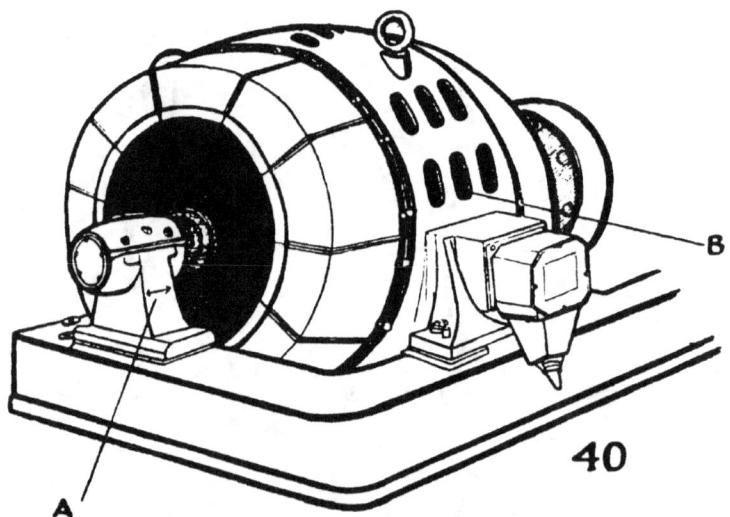

ALTERNATING CURRENT MOTOR.—If the machine is running, place Charge(s) at A. If the machine is stationary, place Charge(s) through holes B.

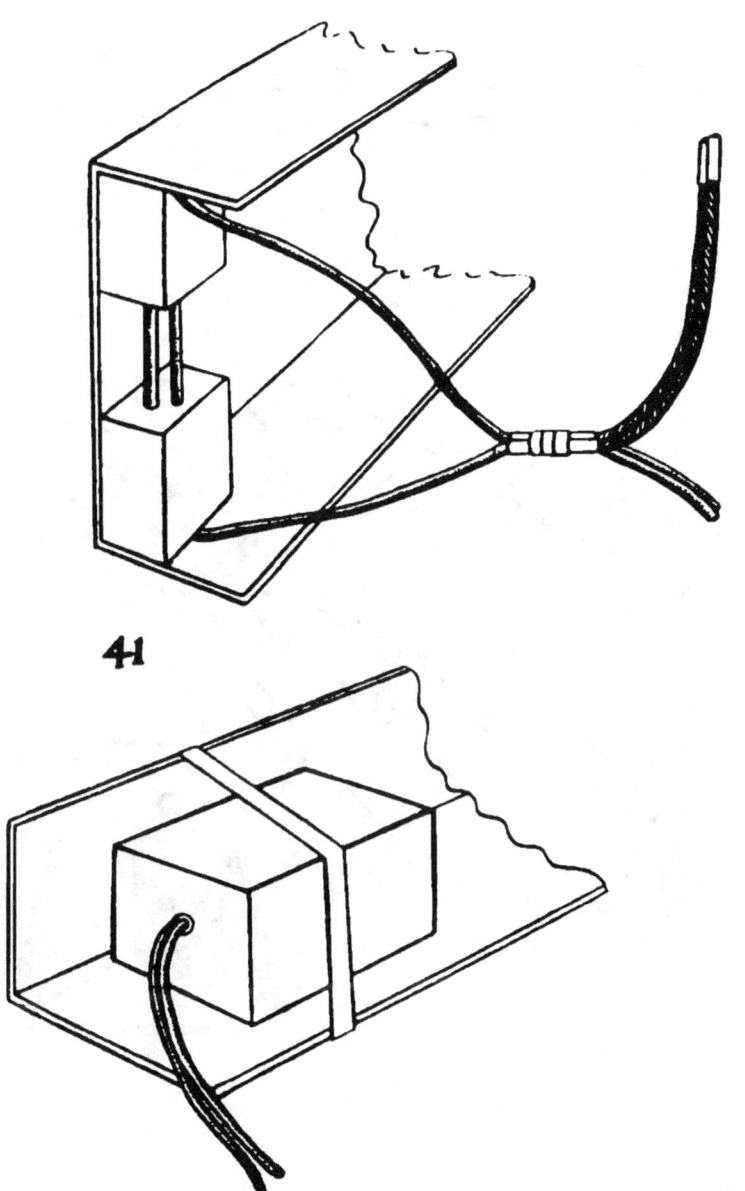

41

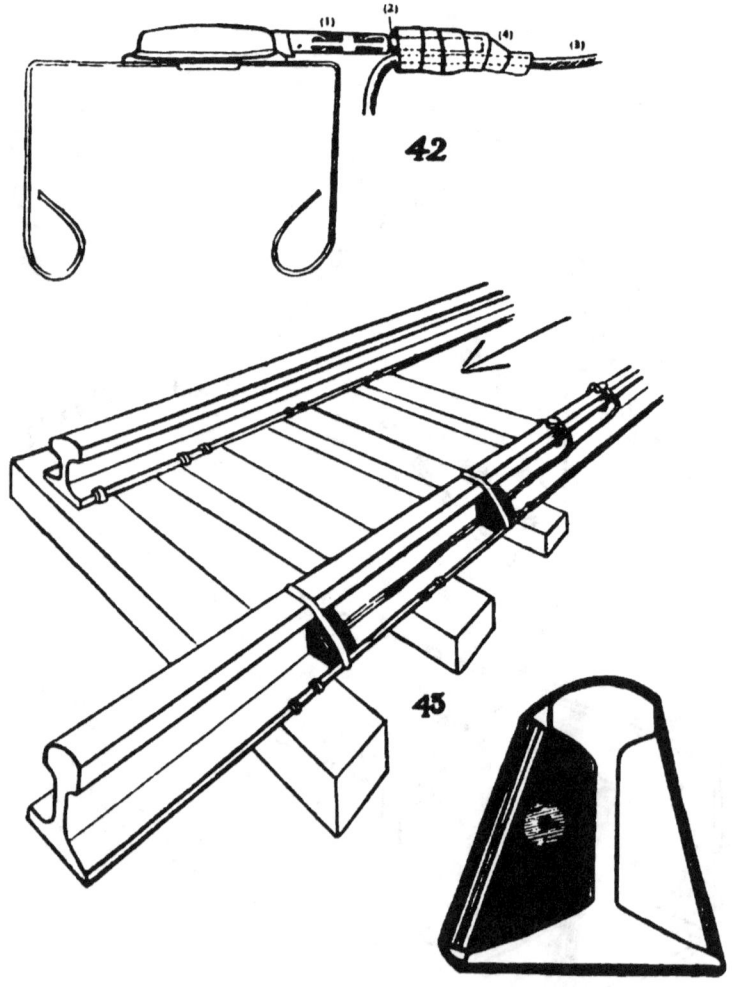

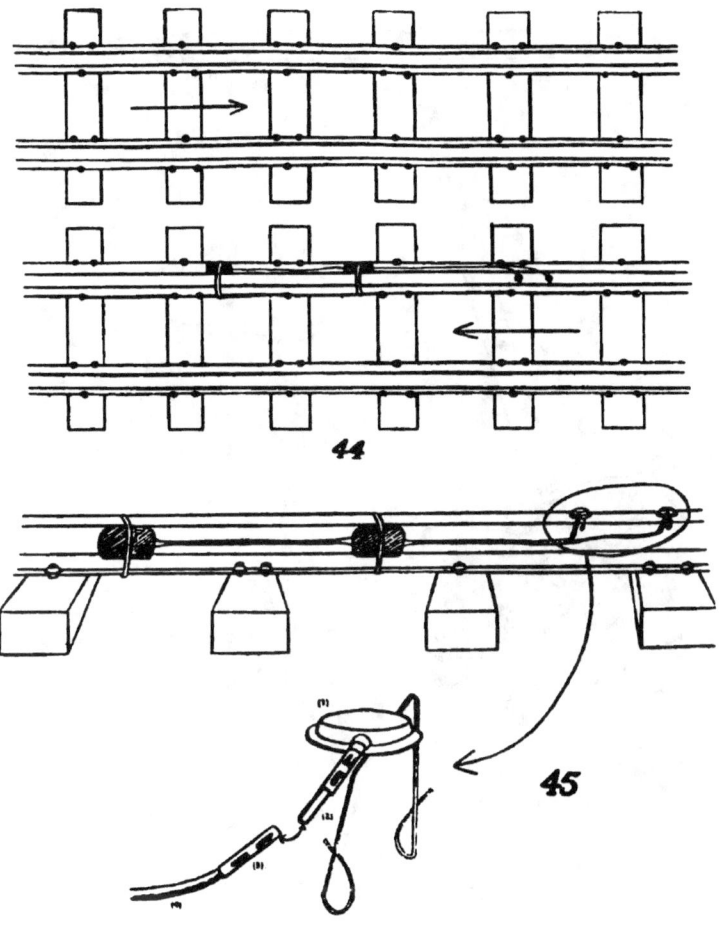

44

45

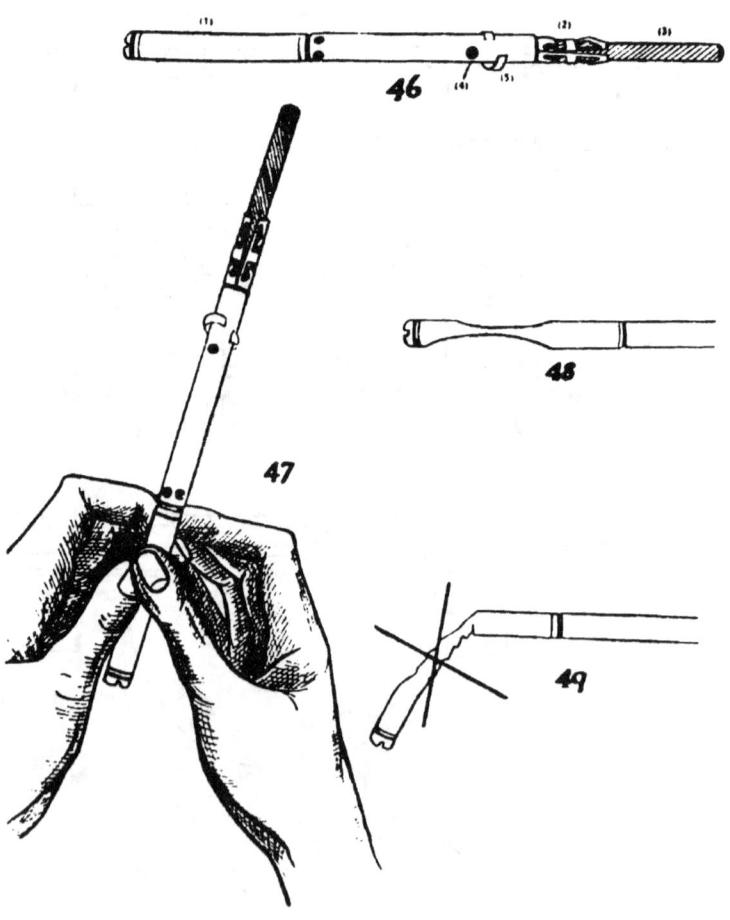

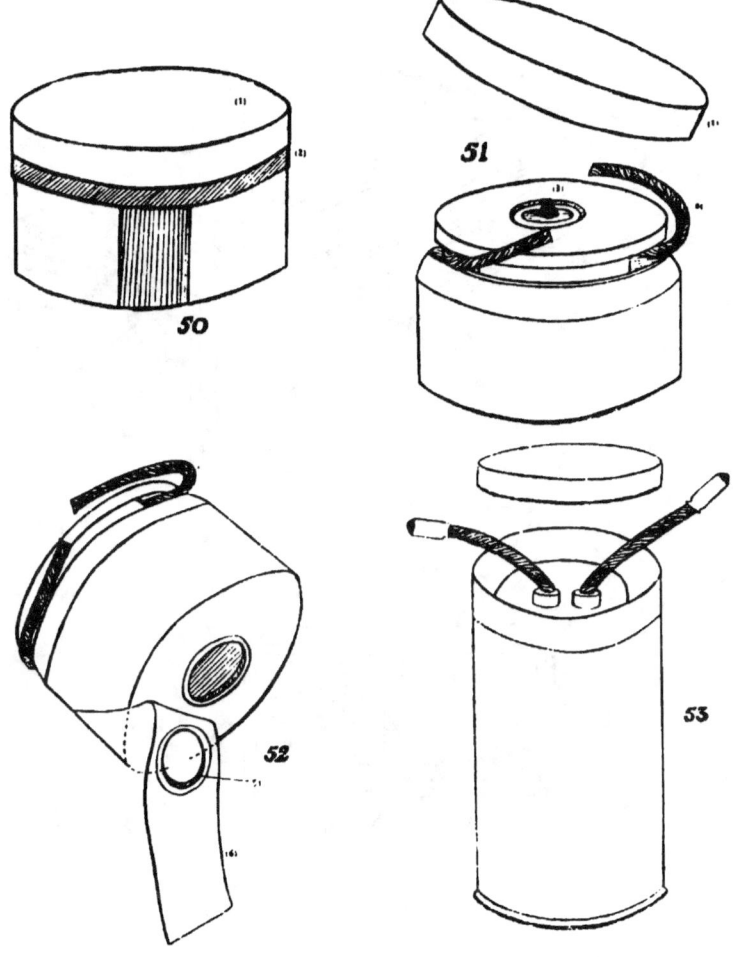

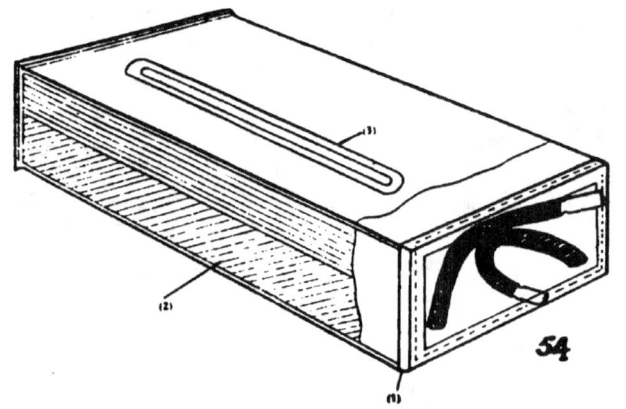

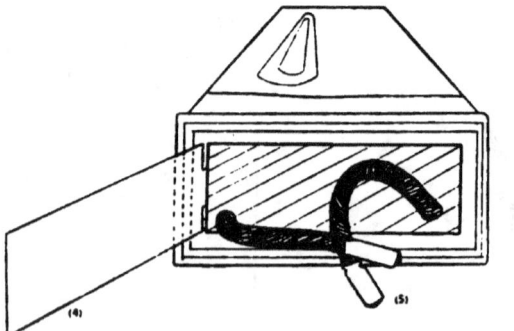

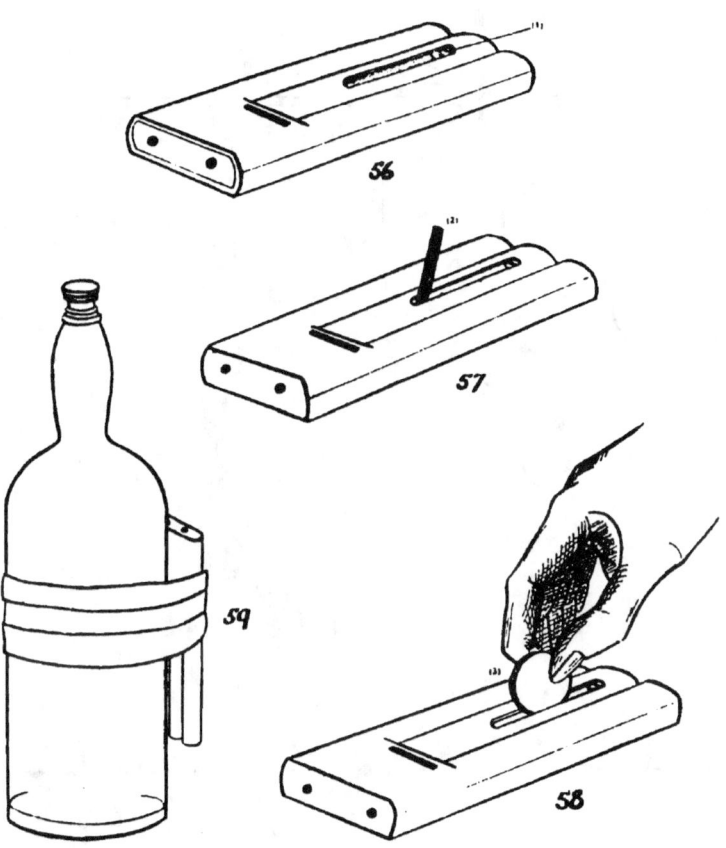

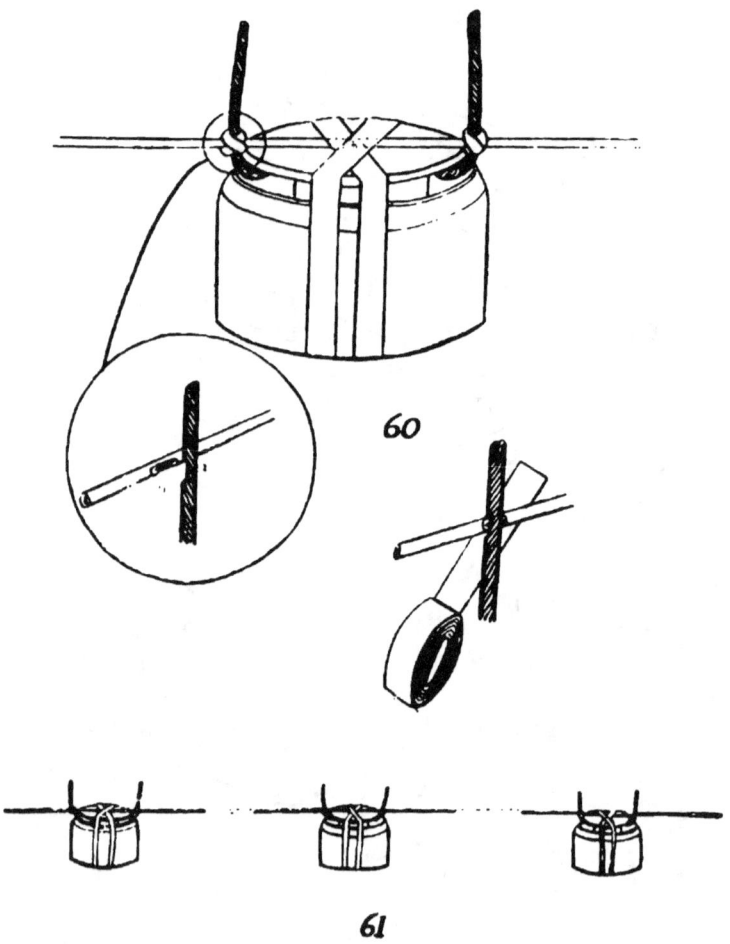

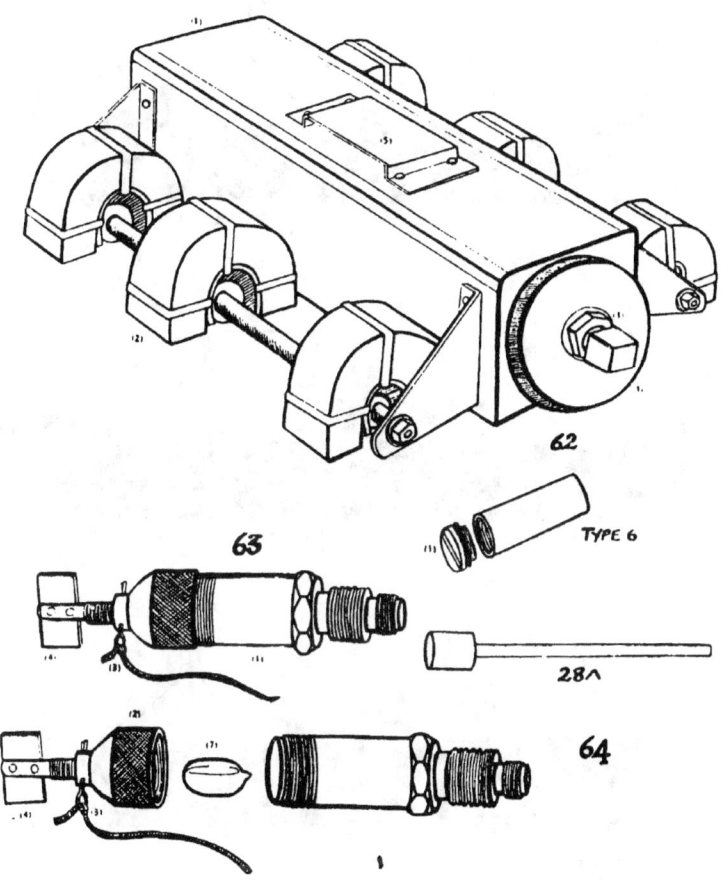

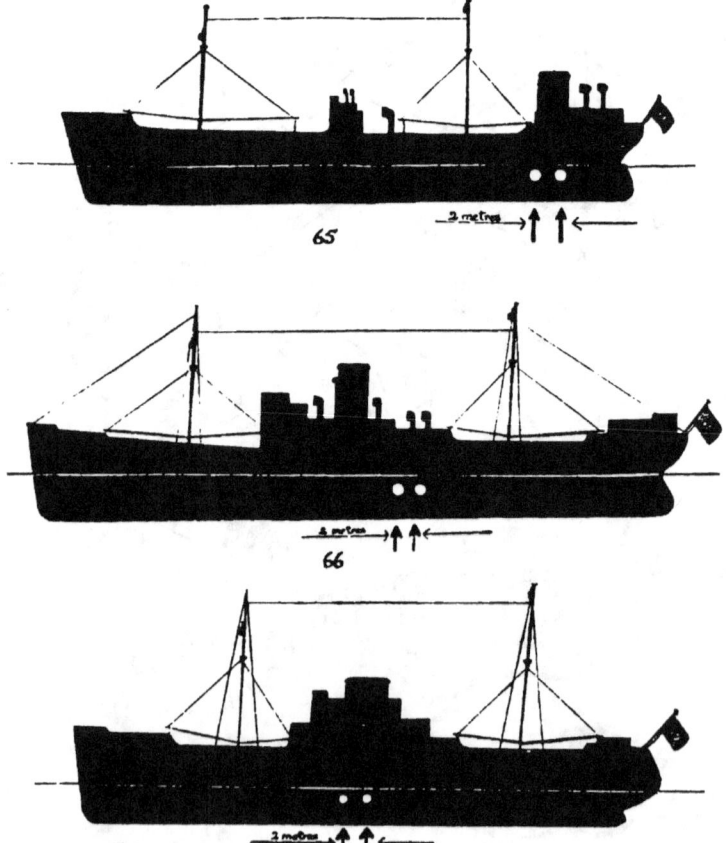

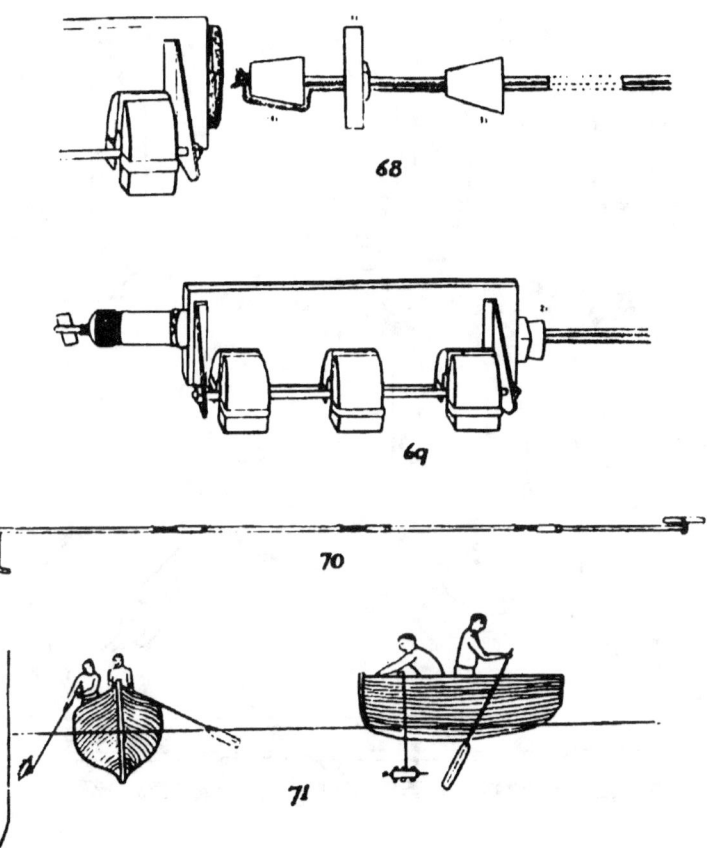

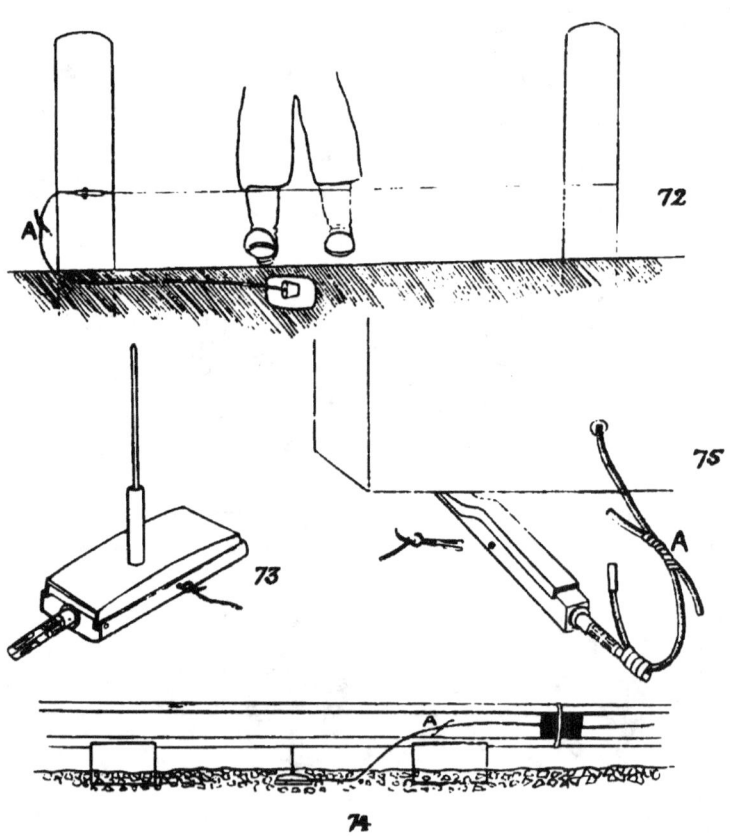

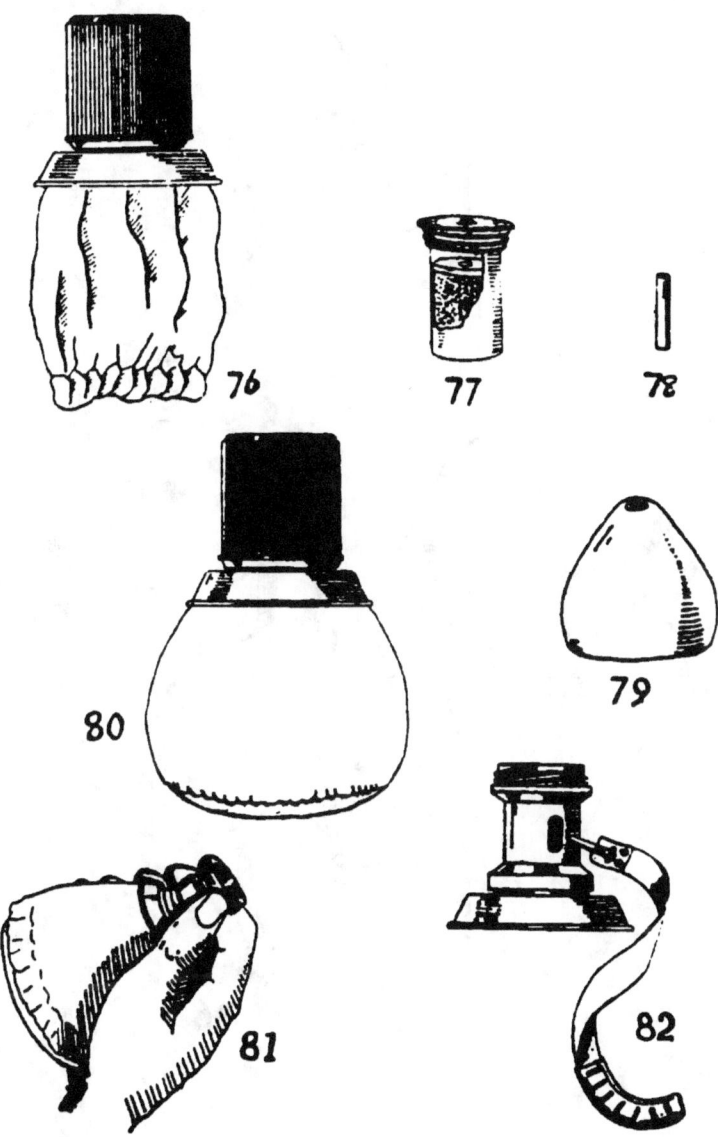

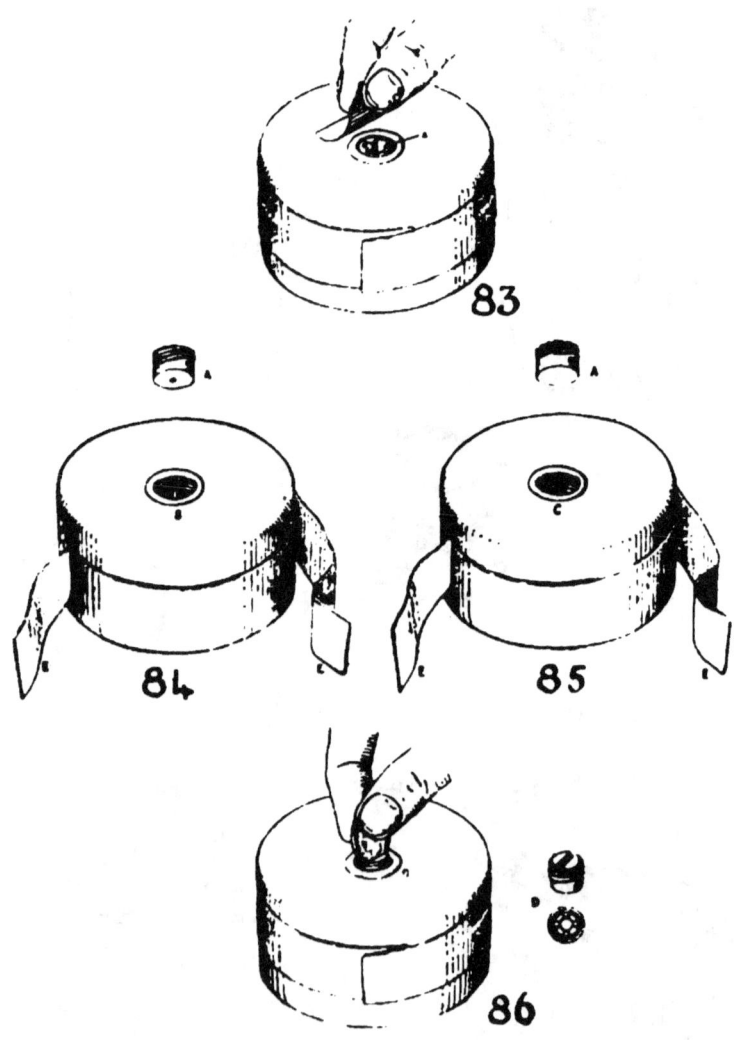

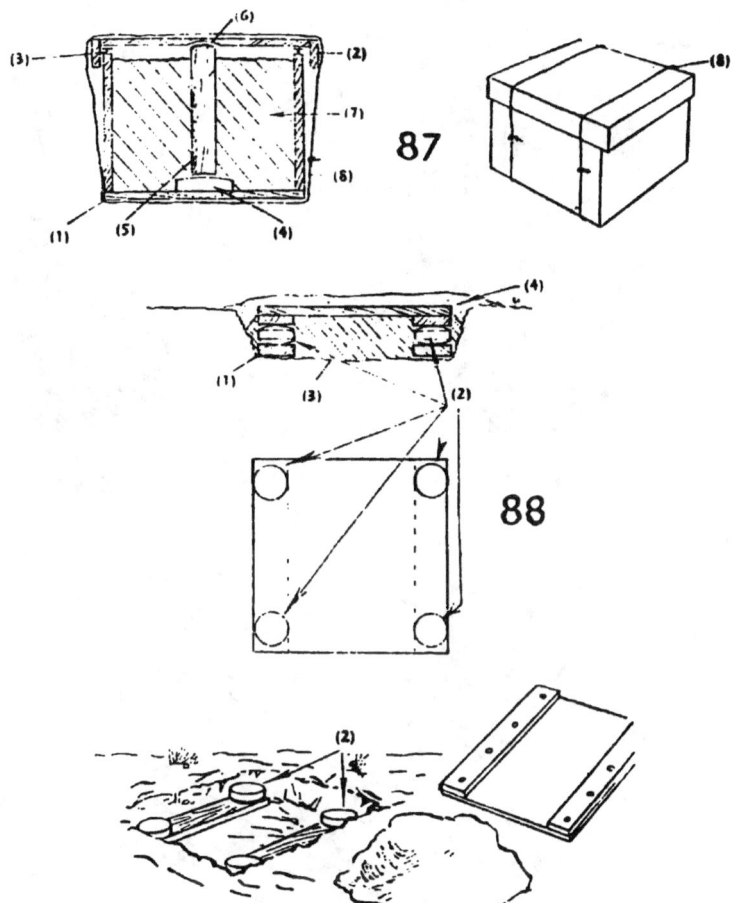

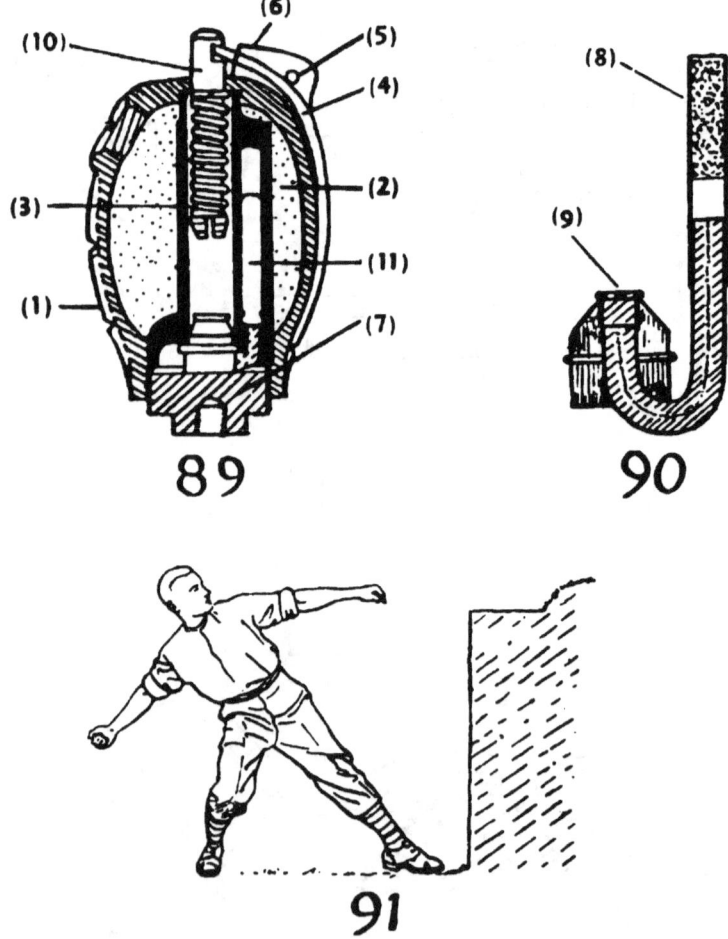

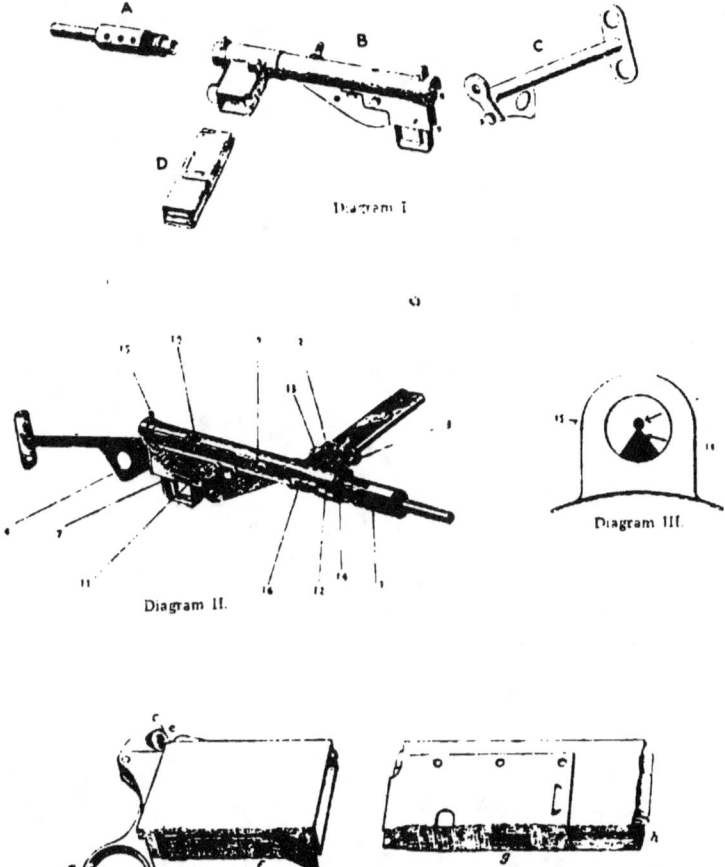

Diagram I.

Diagram II.

Diagram III.

Diagram IV.